Psycles

Dwight H. Bulkley

Psycles *Using Your Circadian Rhythms to Control Accidents, Illnesses, and Psychological Problems*

THE BOBBS-MERRILL COMPANY, INC. INDIANAPOLIS/NEW YORK

Published by The Bobbs-Merrill Company, Inc.
Indianapolis New York

Designed by Mary A. Brown

Manufactured in the United States of America

First printing

Library of Congress Cataloging in Publication Data

Bulkley, Dwight H
Psycles.
1. Success. 2. Occult sciences. I. Title.
BJ1611.2.B84 131 80-692
ISBN 0-672-52651-4

To the thousands of persons
whose sufferings helped to forge
a new path to end suffering.

Contents

Preface

ONE basic discovery—so obvious it hardly needs proof—explodes a thousand mysteries of human suffering and promises enormous practical benefits for all of us.

Called Psycles, and based on a set of exact time-lags which are universal for all life responding to stress, it lays bare the hidden reasons for an endless variety of ills and accidents.

From itches to heart attacks, from hives to hypoglycemia, and from stubbed toes to car wrecks, these and many other mishaps turn out to be precise and intentional responses to something which happened at such times as 37 hours before. They don't "just happen." Nor do they happen just any time. These reaction cycles are specific. And because there have always been these cause-and-effect time-lapses between the triggering-stress incidents and our uniquely logical responses to them, it follows that we have been looking at the wrong times and wrong places for the real reasons for most of our acute troubles. Who would have thought of looking back exactly a day and a half, for example, for the cause of some painful event today?

Knowing these time-lags, and how to backtrack to the specific trigger incidents underlying acute troubles, brings to an end 5,000 years of poor guesses and bad theories on the causes and cures for a large area of human suffering. We are on the threshold of one of the biggest upheavals in the history of science. By pinpointing the deeper roots of accidents, diseases, and behavioral troubles as never before, Psycles opens up a revolutionary new path toward the fulfillment of our dreams of health and happiness.

Although it is only now entering the experimental stage of scientific research, we do not need to wait the many years of "proof" before putting it to practical use. Already its 2,000 case histories provide a new and massive evidence that we actually control essentially every cell of our bodies, and are vastly more aware, subconsciously, than we give ourselves credit.

This new data indicates that there may be few real "accidents," that we modify and distort our tissues and cells on purpose, and that we actually engineer most of our painful troubles with magnificent precision.

But more importantly, Psycles is like a Master-key to the unconscious, a key which automatically unlocks the root causes so that we can literally solve a great many medical and psychological mysteries to bring about healing and to free ourselves from a great deal of needless suffering.

We are thus entering a new era, a turning point in history. No longer can we claim innocence for our mishaps and ills. We have been restored to our central birthright as masters of our own destinies; we have been given a new key to freedom. In due time the health sciences will see their hodgepodge theories on disease and healing give way to this new data. Psycles will alter all our lives. Its time has come. *Our* time has come.

Acknowledgments

To the marvel of how everything happens so rightly and easily when its time has come.

After the vicissitudes of some twenty-eight years of work on *Psycles*, I journeyed to Hawaii to visit with my surfing son Brian, was invited to give a seminar at a health convention, met Carla Annette and all her friends, who introduced me to author Robert Stone, who referred me to his equally marvelous literary agent Morton Dorchin, who went to New York and secured a "yes" from the courageous publisher Jim Fisher and executive editor Grace Shaw of Bobbs-Merrill, who then assigned a bright and perceptive editor, Lynne Spaulding, to *Psycles*—

And suddenly it's all happening!

To all the many friends with whom I have dialogued during the years of shaping this book and who shared or endured my sometimes exasperating probings into the mysteries of human behavior:

Hi Almond, Howard Alltounian, Gordon Beckstead, Arvard Belden, Hjal Berg, Hamid Bey, Dick Biow, Jim Blossey, Leonard Breger, Wilma Bronky, Ralph Bruksos, Roy & Jimmie Bucknell, Dan Bulkley, Spencer Bulkley, Liz Burrage, Billie Burton, Jean Chafets, Perry Chapdelaine, Virginia Criley, Norma Davidson,

Jack Denninger, Edward Dewey, Nadine and Alan Downing, Ryan Drum, Ede Ellison, John Gilbert, Bob Gill, Adolph & Pauline Gottfurcht, Harvey & Rachael Gragg, George & Joan Griffin, George Harper, Bob & Joi Haynes, Lew Hopkins, Jack Horner, Eph & Joyce Howard, Ronald Howes, Ralph Hyatt, Ed & Suni Jacobs, Lou Keary, Timothy Leary, Mamie Lilly, Doris Lund, Pat Magik, Ron Martinez, Jim Morrisett, Rob Mueller, Harry Murphy, Bill Nelson, Barbara Pedersen, Bill & Rae Perrier, Beth Pewther, Ward Phillips, Phil & Sue Rand, Peter Rabe, Don & Carla Redington, Gale & Ann Reedy, Natasha Renard, Bob & Myrtle Robinson, Sheri Shaffer, Jean Sieverling, Robert Smith, Jack & Mary Stanton, Paul & Karen Steinback, George Talbott, Kuni Teramura, John & Diana Thompson, Nancy Tippett, Toni Tipton, Miriam Wilson, Frank & Barbara Wright, Andy & Kandi Youngstrom.

To all my church friends and extended family at the Solo Center in Seattle, who participated in these adventures of the mind:

Al, Alice, Ann, Barbara, Betsy, Betty, Bill, Bob, Bobby, Bonnie, Carol, Chris, Claire, Clayton, Connie, Cindy, Cy, Dale, Dan, Dave, Deedee, Diana, Diane, Dick, Doug, Don, Dorothy, Ed, Ellie, Floyd, Fonda, Frank, Fred, Gail, George, Grant, Howard, Irene, Irv, Jack, Jenine, Jan, Jane, Jean, Jerry, Jim, Joan, Joanne, Jo, Joe, John, Joyce, June, Karen, Kathy, Karol, Kay, Ken, Larry, Lee, LeeAnne, Leslie, Lorna, Lewis, Lorraine, Lois, Linda, Liz, Lynn, Marge, Marion, Mariel, Marilyn, Mary, Mary Kay, Mike, Nick, Norman, Pat, Paul, Pete, Rich, Rick, Robert, Rosie, Rupert, Ruth, Sharon, Shirley, Sig, Stacey, Steve, Sue, Susan, Ted, Tom, Val, Vera, Virginia, Walt, Wayne, Wendy, Will, Yolanda.

To magazine editor Donna Meadors for her years of encouragement, and finally to Baldwin and Dorothy Anciaux, whose warm and friendly home I shared while pounding away on a typewriter, dreaming dreams. . . .

My deep gratitude and best wishes in this greatest of adventures.

DWIGHT BULKLEY
Seattle—1980

1 Introduction

PSYCLES: a) A revolutionary means of pinpointing the specific trigger incidents—and hence the subconscious motivation—for many human mishaps, self-injuries, acute psychosomatic symptoms, and behavioral aberrations. Consisting of a predictable set of time-lags (12.4 hours, 37 hours, etc.) between stress or trauma and troubled responses, this derivation from the universal circadian rhythm (biological clock) solves the causation of most acute human suffering. b) Any such time-lags which are early multiples of the circadian half-wave.

WITH the breaking of the "circadian code" has come a revolutionary solution to one of the oldest problems of health and happiness—the precision unlocking of the hidden reasons for a vast spectrum of human suffering.

Down through the centuries the wisest men have said that we determine the events of our lives. Whether it be cancer, hives, strokes, car wrecks, colds, or getting robbed, most, if not all, of our troubles have been declared to be self-willed. We are responsible. And, much to our displeasure, we have been told that we do them on purpose.

We have also been assured, however, that we can quit such troubles and lead a more trouble-free life, if we both recognize this responsibility and make contact with the basic reasons why we bring them about.

Therein lies the big problem.

Since we don't want to hurt ourselves consciously, our self-responsibility must therefore lie in the subconscious. And this means that we must penetrate that barrier, somehow, to uncover the hidden reasons for our suffering, if we wish to be healthier and happier.

How can we do it?

Traditionally, this has been extremely difficult and time-consuming, even with the best of existent methods. All therapists can attest to that. We resist vigorously; our defenses against remembering the pain of the past are enormous and powerful. Of course we want to change, and of course we want to be healthy and happy, but we avoid these root contacts with a passion, and are apt to continue doing the self-defeating and self-destructive things in spite of our good intentions. Solving our subconscious problems has never been easy.

How, then, can we hope to escape troubles and avoid suffering—if it is so difficult to penetrate, and counteract, the root motives in the subconscious?

From a most unexpected source—the famous and universal circadian rhythm of biology—has come a new and revolutionary solution to this problem. Because of a predictable set of time-lags built into that biological clock, we now have a bridging mechanism between the conscious and the subconscious, whereby we can quickly penetrate to the hidden roots of a thousand painful mysteries, to solve them.

Although most of our responses to environmental stimuli are in present time, there is another large category of stimulus-response behavior in which there is a lengthy, but predictable, time-lapse between the incidents which upset us and our corrective responses to them. The timing of these delayed reactions is circadian. And with the breaking of the circadian code we unfold the secret causes of an incredible variety of human suffering. The subconscious opens up like a flower.

Instead of our saying that accidents or acute symptoms "just happen" without rhyme or reason, we can now pinpoint, as never before, the hidden reasons for many such troubles as car wrecks, hives, cancer, strokes, or robberies. By using the knowledge of specific time-lapses between stimuli and responses, we can backtrack from the acute troubles to their specific causes, to reveal the reasons why we did them. What we find is that they were usually purposive, to correct upsets which may have happened as much as 37 hours or three days before. And with knowledge of this

Psycles timing, we can sometimes solve serious problems within minutes, for therapeutic insight and healing.

After 28 years of investigations, and some 2,000 case histories covering a wide range of suffering and mishaps, we have come to the inescapable conclusion that, far from being the innocent victims of chance forces "out there," for the most part we actually engineer both internal and external troubles with the most fantastic precision—just as those ancient sages said.

Psycles is an exciting discovery which will have a profound impact on the healing arts and the health sciences, on criminal justice, on education, social welfare, and the arts. With it we have a marvelous illumination on many of the most ancient problems of life, of suffering and death, and it beckons us toward a new gateway to freedom.

Three Examples of How Psycles Solve Problems

As the quickest introduction to the whole subject of Psycles, and to show how we can breach the subconscious barrier for insight and healing, we will tell three stories. Taken from among the 2,000 case histories on file, each is typical in illustrating one of the three broad categories of human troubles—accidents, psychosomatic symptoms, and mis-behaviors. Prior to Psycles, they might all have been explained as accidents of nature, to persons who were either innocent victims of circumstances or necessarily being punished for errors in past lives. But with Psycles, none of these explanations appear to be true.

A. Accidents—and the Flaming Hand

We all know what an "accident" means. A slip of a knife and we cut our finger; a slightly wrong motion, and we hit our thumb with a hammer; we collide with someone we didn't mean to; we make a misjudgement or aren't careful with the proper attention and we have a collision with the car; we lose our balance and fall; we grab a rose stem wrongly and get jabbed with a thorn; we fumble our grip on something and it falls and is broken; we spill, slip, slop, and break things, and may even damage our bodies at the same time. These are all readily understood as "mishaps" or "accidents." And we continually ask the age-old question: why did that have to happen?

In a recent college seminar, the students were asked whether anyone had had an acute trouble or accident of any kind within the past day or two. One young man responded as follows: Yesterday afternoon I stopped on the highway to help a stalled motorist. Eventually, I was pouring gasoline into an open carburetor, when a flash-back enveloped my right hand in flames, causing these burns.

When asked what happened Monday morning at 3:30 A.M., (37 hours before the incident), he said that the baby was crying nonstop. His young wife had tried everything and failed to pacify the infant. So, in an unthinking moment, he got up and spanked the baby in the crib. As he returned to bed, his wife cried, "You didn't have to hit her *that* hard!" The same "guilty" right hand was intentionally burned 37 hours later, in what was apparently a deliberate self-punishment for something he instantly wished he hadn't done. The circadian timing is the giveaway.

B. Psychosomatic Ills—and the Throat Hemorrhage

After days or weeks of reasonably good health, we suddenly come down with the flu, or begin to feel "sick," get headaches, stomach upsets. And we wonder why. There are a thousand different symptoms that suddenly happen, and down we go in pain and discomfort. We are told that we "caught" something bad, and we are the "victim" of some "disease." Things go wrong with the body structures or functions; suddenly we have hives, or hepatitis, or asthma, or heart palpitations, or a thousand other aches or pains. And for them there are a thousand different explanations. Many of these deviations from the healthy state can be solved with Psycles.

Eighteen-year-old Judy started spitting blood on a Sunday morning at about 10.00 A.M. When asked what happened Friday evening at about 9:00 P.M. (37 hours prior), she burst into tears, describing a terrible scene with her mother. She had been totally rejected as being worthless, hopeless, "no damn good" by a mother who had "given up on her." With this recall the hemorrhaging stopped.

Searching the deeper meaning of the symptoms, she was asked for an early childhood memory which might have similar elements. She recalled that she had almost died of a strep throat at age three. Hemorrhaging profusely, she was told by her doctor that he had done all he could (this was before the days of antibiotics), and that the rest was up to her. That night her mother was at her bedside all night long praying for her to live. Deep in her heart was that equation—connecting the bleeding throat with total mother love. So, years later, 37 hours after feeling total rejection, she produced the bleeding in her throat because of its associated meaning—love.

C. Mis-Behaviors—and the Crying Jag

It is a common experience for all of us suddenly to do something which we would not, under ordinary circumstances, have done. It can be embarrassing. On impulse, and without premeditation, we find ourselves getting into conflicts, fights, and arguments with people. There are sudden mood changes, depression, anxieties, or sharp fears. We lose things or get lost. There may come weird hallucinations, or nightmares, or fears of going crazy, and we may even feel like ending it all with suicide. All of these nonoptimum behaviors and feelings may occur without an obvious cause in the immediate situation. Why? Let's see how Psycles solved a typical one.

A young and voluptuous female store detective, going about her business on a Sunday morning in a department store, suddenly felt an overwhelming urge to cry. There was no reason for it. She disappeared into a secret viewing hideout called the "towers," which had one-way mirrors for watching customers, and proceeded to cry uncontrollably. She was mystified by the whole event.

Asked what happened Friday evening about 10 P.M., she said that she had a date with a new male acquaintance who seemed a likely marriage prospect. But as the evening progressed, it soon became quite clear that his singular and consuming interest was in her body, not her mind. That discovery was profoundly distressing. She wanted him to be interested in her as a whole human being, not just as a sex object. About 37 hours later, her deeper pain erupted.

These three cases illustrate a monumental breakthrough which is changing the nature of psychology and medicine. In each of these three cases there was the *same* mysterious time-lag of about 37 hours between the traumatic stimulus and the troubled response. Surely there must be a biological basis for this?

Biological Clocks and the Circadian Rhythm

In studying the rhythms of life, researchers have found one central "biological clock" which governs basic living processes from cell chemistry on up. Whether in the smallest single-cell animals, in plants, or in man, it is called the "circadian rhythm," from the Latin words for "about one day," because its frequency is usually a little longer, though varying around the 24-hour solar, or diurnal, rhythm.

Among the 5,000 published items on the subject, there are reports of scores of living phenomena (behavioral, physiological, or biochemical) which manifest this rhythm. They include body temperature, liver enzyme activities, cellular mitoses, respiration, blood sugar, urine secretion, steroid levels, seed germination, skin pigments, running activities, color changes, and photosynthetic rates, to name but a few. Within a single cell, there may be as many as four different chemical events going up and down at different times, but each circadian.

If we put plants in a "black box" with completely controlled environment, their leaves rise and fall, and their petals open and close, in a circadian rhythm, without any known external cues. And if we were to live in a cave, without clocks, and allowed to live spontaneously, rising and sleeping naturally without external programming, our daily cycle of activity would tend to creep away from the normal 24-hour rhythm, and become circadian. This same rhythm, with an average timing of about 24.8 hours (the lunar day), would also be detectable in our internal metabolic functions. Studies of such "free-running" sleep-wake rhythms have been undertaken by Dr. Elliot D. Weitzman at the Laboratory of Human Chronophysiology in the Montfiore Hospital of the Albert Einstein College of Medicine In New York City and by Dr. Jurgen Aschoff at the Max Planck Institute in Germany.

And, finally, this phenomenon is the basis for the so-called jet-lag problem with world travelers, who often find their internal rhythms thrown uncomfortably out of balance in the new day-night location. Sometimes it takes a few days to get back into the swing of things.

So much for the elementary facts abut the circadian rhythm, which permeates all life. Despite the enormous research thus far, it still remains essentially mysterious, with a host of unanswered questions about its origins. No one knows its exact genesis, and there are profound disagreements among biologists whether it is intrinsic or extrinsic. On the one hand, it may be a function of internal chemistry and the kinetics of key reaction rates and feedback mechanisms. On the other hand, life may be so "tuned" to cosmic forces that these rhythms are a manifestation of geophysical or electromagnetic forces, arising from the rhythmic motions of the earth, sun, and moon. Suffice to say that it is fundamental to life, and shows up in every living cell.

We need not concern ourselves at this time with the various theories formulated to explain the circadian rhythm. For our purposes it is sufficient to point to it as a logical basis for the peculiar 12.5-hour and 37-hour timing which we have found in our nonoptimum responses to stress. Likewise, we need not concern ourselves with the research going on in this field, as most of it deals with efforts to modify, control, or "entrain" the rhythm by superimposing various light-dark or temperature cycles. Little of the work being done relates directly to what we are presenting in this book about ways of solving human problems.

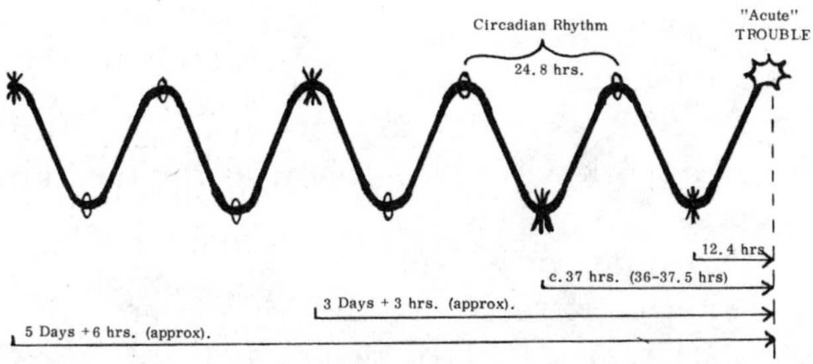

PSYCLES: Many human problems turn out to be delayed reactions to earlier incidents of stress. This picture illustrates the series of stimulus-response time-lags which are universal for all life. This book describes how to use this knowledge to solve problems and guide you to greater health and happiness.

The Discovery of Its Half-Wave

For the most part, researchers in the field of chronobiology were preoccupied with repetitious phenomena, and failed to recognize the significance of events at the *mid-points* of the cycles. Using the analogy of the wave, they studied the peaks but not the troughs. It was as if we had to discover the inverse function of the rhythm, the in-betweens, before the "circadian code" could be broken.

The first occasion when this unique time-lapse was noticed, was during private consultations early in 1952. Acute psychosomatic knee pains occurred suddenly in a patient 37 hours after a therapeutic session involving hypnotic regression into an old knee operation. With this peculiar timing as the initial referent, over a dozen other acute symptomatic events very soon "opened up" in similar fashion, to reveal unique and meaningful trigger causes at the same earlier times. With my associate, Dr. Gordon Beckstead, we then published the original report at the Psychological Research Foundation, Inc. of Phoenix, Arizona, in October 1952.

A month later we received the first confirmation of our observations in a set of 15 similar case histories from the Phillips Medical Clinic in St. Louis, Missouri. A random series of patients were asked to recount events occurring about 37 hours prior to the onset of symptoms or accidents. A unique set of crises, upsets, and serious emotional clashes were uncovered. In several of the cases submitted to our foundation for publication, the insight therefrom resulted in a sudden termination of the symptoms, and a near-instant healing of the problem for which they had come to the clinic.

The main task, in the 29 years since then, has been the patient accumulation of over 2,000 case histories from across the country (and from as far away as India), to promote a solid foundation for the experimental work which will follow as the next phase of the scientific process.

During this time there has been input from Dr. Gardner Murphy, Director of Research of the Menninger Foundation in Topeka, Kansas; Dr. Fritz Perls of the Esalen Institute in California; Father Potter, Director of the Colorado Springs Hospital; and Dr. Elton S. Cook, Director of Research of the Divi Thomae Institute of Cincinnati.

Independent studies were undertaken at the DuPont Company in Delaware, and at the Institute of Humanics in Colorado Springs, where detailed hour-by-hour diaries were kept for four months to test and rule out the significance of other than circadian timing. There were collegiate symposia and a psychological colloquium at Western Washington State College in Bellingham, and a series of radio dialogues which were aired for three years over some 350 stations. All this has led to widespread interest in this amazing phenomenon.

What Sort of Troubles Are We Talking About?

During our 29 years of research we have turned up almost every conceivable kind of suffering exemplifying this phenomenon. They include the typical "accidents," injuries, breaking or losing things, emotional upsets and conflicts, sudden behavioral aberrations, and an enormous variety of illnesses and crimes.

Most of them would seem to involve an "innocent victim" loudly denying his responsibility for the trouble. But because of the same circadian timing between these "mishaps" and their meaningful causes at predictable times earlier, we are forced to conclude that they were all *responses* to the earlier upset; they were done on purpose, were corrective in nature, and uniquely correct in terms of each person's subconscious conditioning.

The essence of Psycles, then, consists of an upsetting experience or incident, a lengthy but predictable circadian time-lag, and finally a specific response.

The following *responses* are typical of our 2,000 cases on file.

State of shock	Stomach ache	Conjunctivitis
Burst ear drum	Psychosis	Knife cut
Throat hemorrhage	Hives	Heart attack
Flaming hand	Car wreck	Colds
Robbery	Traffic ticket	Asthmatic attack
Bee sting	Twisted ankle	Headache
Acute depression	Hypoglycemia	Sudden death

Connecting these responses with their prior upsetting stimulus revealed their subconscious motivation! None of them "just happened." All were apparently done on purpose, as we shall see, to demonstrate that we human beings are truly marvelous in the precision and perfection with which we do our troubles!

That purposeful timing can be our Master-key to freedom.

What Is Meant by Unlocking, Solving, or Curing Troubles?

By unlocking troubles we mean that the use of Psycles (the circadian half-wave) almost invariably opens up the specific trigger-events which resulted in the specific troubles. It "unlocks" by bridging the time-gap barrier between the conscious and unconscious and hence pointing squarely to the subconscious reasons why we bring about such troubles. It points to incidents we might otherwise ignore, repress, or deny as related. It connects the real cause with the delayed response. A thousand painful mysteries open up to reveal their genesis and their resolution.

A problem cannot really be "solved" without first "unlocking" the root causes. If it is merely opposed by counteracting drugs and turned off—as with much of modern medicine—the roots will remain. The difference with Psycles is that we can uncover the secret roots, the deep motivation, the real "whys," and this can lead to solutions and cures in a much deeper and more permanent sense. Only when we have gained insight into the *why* of a trouble are we on our way to being permanently free of it.

One thing we don't mean is instant salvation. Sometimes the results are sudden and dramatic. But more often the process of solving serious problems may take a series of insights resulting in alleviation of symptoms—before a final "cure."

Miraculous Things?

By saying that the use of Psycles can produce "miracles," we don't mean to imply the invoking of divine providence or the supernatural. We mean that the right connection between the trouble and its true unconscious cause can result in sudden and marvelous changes of behavior and symptoms.

Imagine someone having a heart attack, or depression, or suicidal impulse, or any one of a hundred other sudden troubles. Then picture yourself having the ability, by asking one or two simple questions, to help that person terminate the problem just as quickly, without the use of any special drugs, machines, or facilities.

Wouldn't that seem marvelous and "miraculous"?

But Can It Be Dangerous?

The use of Psycles is foolproof.

If the trouble is 100 percent "physical," the questions do no harm, little time is lost, and the usual therapeutic procedures can be followed.

But if the problem is "psychosomatic," which most authorities agree represents from 70 to 90 percent of all problems, then Psycles not only does no harm, but can result in "miraculous" transformation, alleviation of pain and anxiety, or at least some partial insight into the whys of the trouble, and a lessening of its tension.

Of course Psycles can also be used in conjunction with various chemicals which oppose the painful condition. In addition to the helpful action of the drug, then, the insight from Psycles can be immensely beneficial in the deeper sense, in terms of the whole personality and its motivation for creating the trouble. It deconditions and liberates, quantitatively.

Is Psycles Another Panacea?

No, it is a single key to many answers, rather than a single answer. Because human beings are all unique creatures, this single key unlocks thousands of unique answers. Psycles is like a guide, a yardstick, or a timescale filter, which exposes the thousands of unique and different causes for the thousands of specific troubles. The timing is what is universal. What would trigger conjunctivitis for me would not necessarily do so for you. What would trigger nausea for you would not necessarily do so for me. We are all unique and different because of our experiences and associations. Psycles can uncover the unique causes for each of us.

There are many types of chronic troubles to which Psycles does not apply. It applies only to "acute" troubles—those troubles which have just happened, at a time which can be specified. In the case of chronic troubles, however, there are the occasional flare-ups which may be suitable for analysis.

How Simple?

The use of Psycles is so simple that almost anyone can employ this knowledge to bring about healing therapy for themselves or for others in trouble. With this information you can make use of a precision tool which works in a high percentage of cases.

Psycles is easy and understandable, and with it you can do "miracles." We will show you how to use it. The troubles do unlock. They break wide open. It has been called a "karma-buster" by those who believe in retribution for past lives. We will give you many examples of actual cases to show how it works.

And the secret is open to all. With this revolutionary tool you can storm the gates of Heaven—within. Take it. Use it.

Don't believe in those who say that the universe has no simple answers, or that there are no easy solutions to human troubles, or that God has decreed that we should remain ignorant and in suffering. Such beliefs belong to those who claim to be innocent victims, standing by in a hostile universe which jabs troubles *at* you.

Because we *are* responsible for our troubles, and do them on purpose, there is this escape, this Master-key to the unconscious, which can pinpoint for each of us our own way to freedom!

An Invitation to Explore

The invitation is extended, therefore, to explore the many exciting facets of this discovery, with a whole new fresh look at the ways in which we engineer our painful limitations. Because if we intend to change ourselves, and if we really wish to be liberated from internal bondage, the sooner we come face to face with our own fantastic responsibility for our events the better. There would seem to be little justification for going on pretending that we are innocent victims of a capricious god in a hostile universe. The world is full of pretense that gets nowhere.

Beginning with accidents, and then going on to psychosomatics and other categories of human troubles, we will discuss scores of actual cases where we turned the clock back 37 hours (etc.) to find the hands pointing squarely at underlying causes, which provided precision insight for their solution. We will also develop some more specific things which we can do, as an ongoing process in utilizing this information to promote our continuous progress in the great adventure of self-liberation.

For we are not just a bundle of atoms, bouncing around by blind chance in a hostile universe. With this Master-key we can be restored to our central role in life and once again discover how we determine events, how we can unlock our binding troubles, and advance with new insight to a more trouble-free life.

2 Accidents

ONE of the roadblocks to understanding accidents is that we have all been so brainwashed on the subject. From earliest infancy we have been taught that an accident was an accident. We didn't mean to do it. . . . We didn't want it to happen. . . . It was meaningless and just by chance . . . wasn't it?

But as we use Psycles to penetrate the surface of such events, the entire definition begins to crumble. Instead, we find people doing "accidents" with precision, on purpose and for necessary reasons. For the most part they are not chance events, but wanted, and done for equilibrium purposes. Seen in this light, they become examples of what is known in biology as homeostasis.

Let's sidetrack briefly on this subject.

Understanding Homeostasis

All living beings, from single cells to man, actively maintain a balance in the preservation of their own integrity and continuity. This is called homeostasis. Any stimulus can be considered as a disturbance to the equilibrium of the organism, and every response, then, can be visualized as an effort to restore equilibrium.

An amoeba which avoids bright light is manifesting homeostasis, just as we are when we withdraw our hand from a fire, or our flesh heals from an injury, or our faces blush to a complimentary remark. Many physiological and psychological changes serve to maintain the continuity of a living structure.

The changes which we most commonly see happen immediately: the recoil from a slap, the cessation of hunger pangs after eating, the speeded-up heartbeat when danger threatens, the dilation of pupils in darkness, or the clotting of blood in an open wound. While we are driving a car, we are continually responding to an endless series of visual signals, signs, sounds, and other sensations. All of these more elementary responses follow very soon after the stimuli.

But now, with Psycles, we are discovering a whole new field of responses which involve distinct time-lags. They are still homeostatic, but they are quite different from the ordinary ones. They don't happen immediately following their stimuli, but occur only after a lengthy, but predictable, period of time. Furthermore, these time-lapses between stimulus and response, or between cause and effect, are uniquely circadian, because their intervals are multiples of its half-wave.

This category of response, which encompasses a wide variety of accidental mishaps, psychosomatic and behavioral ills, apparently involves the ego, or conditional self-image, and its subconscious conflicts. Since these responses are made to maintain equilibrium, they can be labeled as "psychosomatic homeostasis."

We can now take a completely new look at "accidents," viewing them as homeostatic responses which follow their specific motivating stimuli after a unique circadian time-lapse.

As the stories unfold through these pages, you will surely find new and fresh insights for solving some long-standing problems. For in spite of the pretentious illusions that surround us, it *is* a marvelous world, and we don't have to remain blind forever!

WHY WE BREAK THINGS AND HURT OURSELVES

The Grass Whip

One Sunday morning at about 10:30 I was cutting tall weeds behind the house with a grass whip, when I struck an overhead clothesline, deflecting the swing in such a manner as to bring the blade down into my right leg, smashing through the muscle into the shinbone just above the right ankle.

While dancing about on my left leg, leaving a trail of blood, the first thing I did, through my tears, was to look at my watch!

What time was it? And why did I have to do *that* to myself? And then, still in contortions of agony, I did the calculation, back 37 hours to Friday evening about 9—9:30—and . . . *oh!*

There it was! The trigger. The reason. The *why*.

In the kitchen, between 9:00 and 10:00 P.M., Friday about 37 hours before, there was a loud and noisy confrontation with my son. Having just turned 16, he was pressuring me for money for a car. "Right now," I was saying, "we've got some heavy bills, and it just doesn't leave us any spare cash." After a series of point and counterpoint, his final thrust was, "If you're so damn smart, why aren't we rich?"

Basically, he was challenging my self-image as a good father. And I was feeling very defensive, insisting that he was not so deprived.

But why the smashing wound above my right ankle?

It wasn't until several hours after the so-called "accident" that I remembered the key fact which brought meaning into the whole event.

Many years ago, *my father*, who was then a medical missionary, was swimming in the Bay of Bengal, off the southern coast of Thailand, when he was struck by a giant stingray. He managed to make it to shore, but lay unconscious on the beach for hours before he was discovered. For the rest of his life, he carried a large, circular black scar just above his right ankle!

Meaning and motivation began to emerge.

On Friday evening, I was attacked in the middle of my self-image—was shaken, and defensive. And Sunday morning, about 37 hours later, I "accidentally" wounded myself, to produce a mark on my body which was *like my father's*. Without the knowledge of my father's wound, the two events would not seem meaningfully related. But *with* this fact, and with some understanding of the *essence* of what happened between me and my son, the picture changed entirely. Suddenly, there was significance, with new dimensions in insight.

Also, consider the precision with which I handled the grass whip—how I hit the clothesline just so, to deflect it just so, to bring it down just so, to strike the correct spot on my ankle—all subconsciously.

We have hundreds of "accidents" which reveal their motivation through Psycles.

The Broken Vase

On a Thursday noon I was washing some dishes, and in my soapy hands was a green glass vase which I fumbled, let slip, and smashed.

Was it because my hands were slippery? Was I not being careful? Or wasn't I paying attention to what I was doing? Or was my attention distracted? To all these questions, the answer could be a superficial yes—but not really!

As a matter of fact, on Tuesday evening about 37 hours earlier, I was returning from a lecture with a female acquaintance, when we stopped at a cafe for a hamburger and coffee.

She took out her cigarettes and offered me one. Having just resolved to quit the day before, I started to refuse, but was obviously unsure of myself.

"Oh, go ahead and have a cigarette," she snapped. "You guys who try to quit smoking are more damn trouble to others. . . ."

With very little argument, I "weakened," "broke my resolve," and resumed the habit. But deep within me there was anger, apparently, indicated by what I did 37 hours later. I smashed a vase, which, in classical literature, would be considered as a *symbol* of her.

The Reactor Vessel

In a somewhat similar vein is an incident in a chemical laboratory, involving the "accidental" breaking of a reactor vessel. On probing the events 37 hours earlier, it turned out that the researcher and his wife were in conflict over the spaying of their female puppy.

Here again we have what appears symbolic of female structure—the reactor vessel, as a "womb" in which ingredients are mixed and cooked, for a product—and hostility directed against the symbol instead of the real object.

Understanding Our Symbolic Behavior

The study of Psycles confirms how enormously symbolic we are.

If the self-image is threatened in some manner, and if we cannot respond aggressively against the persons themselves right then and there, we may have to deal in similar manner with a *symbol* of that person, Psycles later. The classical male and female symbols, which have been described again and again throughout psychological literature, will be found repeatedly in the circadian analysis of mishaps.

Substituting for a man/husband/father-image, we find such objects as knives, guns, machines, tubes, pipes, and other rodlike or mechanical things taking the brunt of our rejections and hostilities. And substituting for a woman/wife/mother-image, we often find cups, bowls, boxes, houses, cars, and other hollow, womblike structures featured in our aggressive releases.

The extent to which we act symbolically is not generally recognized. There are a host of small, insignificant activities, relating to objects which appear innocent enough, as long as we are not aware of their unique underlying motivation. But when we filter the events through a circadian time-interval viewpoint between the stimuli and responses, then we find ourselves routinely acting in a symbolic manner on a wide variety of symbolic objects.

Our uniqueness should also be emphasized. Not everyone in conflict with a female would smash a vase or break a reactor vessel. Accidents are engineered in accordance with each individual's personality, background associations, identifications, and problems. We select from a wide assortment options which are right for us. Because of this personal specificity, the interpretation must fit the case. Of course, there are many similarities, and we can easily recognize those which apply to us personally, as we explore the underlying processes. It is helpful to make the connection, even though we may as yet be unable to find a clear interpretation.

Two Coffee Spills

An example of an "accident" with less clearly defined interpretation is the case of the secretary who knocked over a full cup of coffee. The brown liquid cascaded over a pile of reports, making it necessary for her to do a great deal of retyping.

Why did she do this? Two nights before the mishap, 37 hours previously, she was sitting in a motion picture theater, annoyed by a male stranger in the next seat who kept jabbing her with his elbow. How do we interpret this?

Similarly, an engineer spilled a cup of coffee all over his desk, but note what triggered his mishap. In his case, a seven-year-old daughter, awakened by a nightmare, crawled into bed and spent the rest of the night between him and his wife.

From many other cases of spilling and accidental slips, knocks, and bumps, it seems that much relates to questions of self-control, how surely and continuously we maintain our self-image, and how this relates to the performance of roles.

The Near Crash

One Saturday evening about 6, two young men were driving a pick-up truck across the desert between Tucson and Phoenix, when they nearly had a head-on collision.

Jim was driving, and Bob was sitting alongside, when they both heard the sound of an explosion. Jim, thinking it was some dove-hunter in the nearby fields, turned his head to the left, responding to what he thought was a shotgun blast. In the meantime, Bob, looking straight ahead, saw that an oncoming truck had blown its left front tire, and was swerving toward them, dead-on, out of control. In the last split second, the other vehicle regained control, swerved back in its lane, and missed them by an inch. All at high speed. Bob was drained white.

About an hour later, while the two were visiting me in Phoenix and recounting their story, they wondered what might happen 37 hours later, and asked me if I had any predictions.

"Frankly," I said, "I haven't the faintest idea what you might do Monday morning, if anything. Certainly seems like a pretty good shock. But whether you do something specific and homeostatic for that incident, depends on what was going on inside of you—in terms of your own whole personality-structure."

"Suppose we just sit it through, and do nothing."

I suggested that, in the meantime, the best thing they could do would be to remember it as vividly as possible, going back through it, and feeling the fright. Then—if they had to do something to balance the books—the adjustment might be less severe, although there could be no way of proving this.

On the following Thursday I received a letter from them.

"Monday morning we decided to just sit through the critical time, from a little after seven. We just sat and thought about the near crash, and nothing happened, until about 7:30 A.M. when Bob reached down to pick up his lunch box. He "accidentally" hooked his foot in the telephone cord and brought the telephone crashing to the floor, with the receiver off the hook.

"We laughed. And wondered. And that's all that happened. What do you make of it?"

"Well," I wrote back, "if you look at it symbolically, the receiver off the hook is like a communication break. Consider the crisis moment. There was Bob, facing death, with his whole body screaming to Jim 'Watch out!' while Jim was looking out in the fields. In that split second, there was a communication problem. Survival depended on it. 'For Crissakes, Jim—that guy's gonna hit us,' but his message didn't get through. So then, 37 hours later, it's as if that receiver off the hook 'explained' the problem. See? Couldn't get through to you because of that . . ."

Their final reply was "Guess what! That goes right to the heart of Bob's whole personality problem. Communication. He's the strong silent type but when he speaks—you'd better listen!" They described how this one clue got them going for hours, back into many events of his life where communication problems were paramount. It was a liberating experience! Just from that one little "accident," with the phone on the floor and its receiver off the hook.

Again, we say that each person is unique.

Ten different people, in ten near wrecks, could do ten different things to manifest their own homeostatic needs. But in each case, for that person, what happens can be of great help, in self-insight, toward that person's own self-liberation.

The Dark Attic

An outstanding example of symbolic behavior happened one day while I was watching a young man retrieving a box from the attic.

He had a chair and ladder in place, with one leg propped against a far wall, raised the trap door to the attic, gave it a half turn, and then slipped and fell, with a great clattering noise, as the door, ladder, and chair all came crashing down on top of him. He wasn't so much injured as frightened by the whole episode.

What happened 37 hours before?

He was in session with a therapist who had begun to probe a new area in his unconscious, which was frightening, and from which they retreated. The attic clearly represented the dark recesses of his unconscious mind and memories, and the event seemed to symbolize the dangers which might lurk in opening up the unknown—up there in his head!

The variety of action and reaction we generate as we go about our individual tightrope walkings seems almost unlimited. Sometimes we seem to punish ourselves for things which we did, but which we wish we hadn't.

The Garbage Volcano

There was the commune guru who was performing a routine chore one morning at about 8:00. It was customary to put the various organic garbage into the blender for fertilizing the vegetable garden.

He turned the switch on, while the cover was off, and the thick black goop erupted like a geyser, depositing filth on the ceiling, over the clean dishes, pots, and pans, and over his coat, pants, and suede shoes.

It took two hours to clean up the catastrophe.

In order to gain some insight into the logic behind the event, we had to go back 37 hours to find him erupting against one of the commune members, giving her a tongue-lashing for her failure to clean up after herself and her child! That was about 7:00 P.M. two evenings before. Going back another 37 hours might have provided insight into the reasons why he had the impulse to lash out against the young mother for being filthy. But that was not explored.

The Salesman in Hot Water

A salesman, feeling restless, decided to have a hot tub bath shortly after midnight in the hopes that this would facilitate sleep. After soaking in the tub, he reached up and grabbed hold of a wall fixture, which was both a soap container and a handle, to pull himself up and out of the hot water. The entire fixture came out of the wall, leaving a gaping hole and a messy repair problem.

Shortly before noon, 37 hours before this mishap, he had received notification from the factory that his next shipment would not arrive until after Christmas. He had promised many customers a delivery in mid-December.

He was in hot water. He immediately phoned the factory representative to find out why the error in his order and shipment date had occurred, only to find that they had made the mistake but that it could not be corrected.

The handle that was to get him out of the hot water failed him.

THE CRAZY WORLD OF CAR TROUBLES

If we were to see a car abandoned by the side of the road, resting on its axles in deep sand, what might we "see" in that scene? Would the evidence from Psycles make us see it differently, adding depth of meaning to an otherwise familiar sight?

First of all, from all our evidence, we can assume that the driver *wanted* to abandon a car stuck in deep sand. Although his report might suggest otherwise, we can now know that he deliberately, though subconsciously, brought it about just as it happened, and that it was necessary for him to do so.

Secondly, recognizing the obvious symbolism in the event, we can "see" that the driver of the vehicle, probably a male, was dramatizing the need to abandon his wife/mother or female-identified person. He perceived that female (whoever she was) as bogged-down, stuck, and immovable. He felt frustration, resentment, and anger toward her. She couldn't be moved. To hell with her; she ought to be dumped!

That, very probably, was his feeling about 37 hours before he managed to do what he did to the car. Specifically and precisely it could be mapped out, if we had a detailed diary of the driver, with his activities, relationships, and troubles during the past week. Such a map would reveal the subconscious need and motivation, and show clearly why he needed to get that car stuck in deep sand—deliberately, and on purpose.

Does this seem so improbable?

Actually, we are talking about something that is quite universal. Having explored such events for the past 29 years with our new Master-key to the subconscious we can say that this is typical of what we are all doing, behind our deadpan faces and pretenses. And furthermore, it is fortunate that we *can* vent our aggression on symbols—rather than on the real persons. Exposing these hidden motivations and realities can most certainly make us healthier and happier.

The Steering Trouble

Let's take one of our many actual cases which show how psychological problems may be dramatized in car troubles.

"For three mornings, now," one of my fellow laboratory workers said, "I have awakened at 4:30 A.M., feeling jumpy. Can't understand why."

"Is something bothering you about the same time each afternoon?"

"Going home," he answered, "in the heavy traffic."

"Are you afraid of heavy traffic?"

"I'm having trouble with my steering gear . . . hard to drive . . . must work on it."

So far so good. Steering troubles each afternoon. Wakes up 12 or 37 hours later feeling the anxiety. But imagine what would happen if he went to a doctor complaining of waking up jittery every morning at 4:30. Without insight into the psychological roots, he might have to undergo thyroid tests; urinalyses; heart, lung, and kidney checks; besides medication to counteract tension.

With an exploration of its circadian timing, however, we find that every afternoon, as quitting time approached, he became apprehensive about driving home. Out on the freeway, struggling with his car, fighting its steering mechanism, he was setting himself up for the early morning jitters. But we can go further than that. It's not enough to say that an erratic or tight steering gear on his car is the real cause of his trouble. What about the person that the car might symbolize?

"OK, OK, so I'm having troubles. I'm not sure my wife is not straying—don't seem to have her under control, or lined up. We may be in for big trouble!"

Now we have the real source, the real genesis, for the car trouble underlying the nervous trouble. His marriage is shaky. And this is one routine example of how we manage, in various subtle and devious ways, to bring about misfunctions in cars, which are symbolic of our personal relationships.

His subconscious was trying to tell him something.

In other words, the fact that the car had an erratic or sticky steering wheel was not simply an accident, nor was it meaningless. It didn't just happen. The sticky steering mechanism was a precise symbol of his problem, which he apparently engineered as a function of his whole drama.

By using Psycles, we went right to the real problem which needed attention—the shaky marriage.

It would be a legitimate question, however, to ask how the owner of that car could possibly be responsible for the faulty mechanism. Couldn't it be just mechanics, wear and tear on machines—physical laws? Although in this case we didn't find out just how he was doing it, physically, we have concluded from the psychological aspects of the case, and its 37-hour timing, that he was somehow causing the steering trouble as part of his unconscious needs.

Another case will show the precision engineering that was involved.

Harvey's Car

On a Friday evening, at 10:20, Harvey slammed a wrench into the pavement, let fly with his boot at the front wheel, bashed the hood shut, and stormed away, abandoning his English sports car in a supermarket parking lot.

"That's it! I've had it!" he bellowed. "After the damndest hour and a half of working on the s.o.b.—it stalls every time I put the clutch in."

Before unraveling the exact manner in which he engineered

the trouble, we will summarize the psychological situation surrounding the case. Harvey was having difficulties. He was dating a divorced woman with two small children. He felt that she was trying to trap him into marrying her, but considered her "half-crazy." For the past three days, there was a significant trouble almost each Psycle (or circadian half-wave), beginning with a Tuesday evening date during which she was depressed and demanded money which he refused. Wednesday evening, a double date, with emotional troubles. Thursday morning, a date broken. Thursday evening, he went to her house to find a baby-sitter who informed him that she was out with another girl. Friday morning, he went to an insurance office regarding damages to his car from a previous accident.

And now, as he described it, Friday evening:

"The car had been running badly all day. Thought it might be the points, so I set them. Still no good. Checked timing with a strobe light . . . slightly off . . . then adjusted the distributor. Had to pull the tachometer cable out of the way, disconnecting it from the generator. I adjusted the distributor back and forth. Finally ended up back where it was . . . points same . . . timing same . . . no better.

"By this time I was getting mad. Tightened everything back up. Had ten minutes of trouble with one damn bolt on the underside of the distributor. The tachometer cable I hooked back onto the generator . . . timing light off . . . down with the hood. It started up fine, but each time I put the clutch in, it stalled. Started it up, clutch in, and it stalled—three times. Now I was real mad. The service station was closing down next door, and they turned off their lights. That's when I figured I had had it! So I just left the s.o.b. car there on the lot."

The next day, feeling a little calmer, Harvey went back to his car.

Evidently, he had reconnected the tachometer cable on the wrong side of a bare radiator brace, so that it was lying one-eighth inch from the coil. Now he discovered that every time he depressed the clutch, the cable touched the coil, thereby causing the shorting and the stalling.

From our new standpoint, we can say that it took real skill to cause that engine to stall every time he put his foot on the clutch. The fact that it was subconscious makes it no less ingenious.

Again, we say that it would not make much sense, *unless* we use this Psycles Master-key, that universal biological rhythm, to reveal the troubled motivation for doing what he did. *Then* it begins to make sense. He needed to vent his anger against a car (instead of the woman), so he went through this hour and a half of tinkering before ending up with an explosive abandoning. He was abandoning *her*.

This case was especially good because we were able to pinpoint exactly how a car trouble could be engineered, to satisfy a psychological need.

The following case, on the other hand, leaves the mechanical process unsolved.

The Overheated Car

On a Friday morning in Cincinnati, just before 8:00, a car suddenly boiled over, blowing geysers of steam. The driver stopped, filled the radiator with water, but another mile down the road witnessed the same billowing steam of an overheated car. As it turned out, the water line to the heater had pulled loose, and was easily fixed.

Normally, one would not look for motivation in such an event. Water lines pull loose and can be reconnected. But the timing of this was peculiar. It happened just 37 hours after an important cross-country phone call to a mother with very high blood pressure. And note the symbolic relationship between high blood pressure and too much heat and pressure of fluids in tubes in a car. If it were not for that 37-hour timing, we could ignore it. But taking it into consideration, along with 2,000 other cases of circadian timing between troubled stimuli and their homeostatic responses, we have to suspect the possibility that it was meaningful, and on purpose.

Now if this trouble were done on purpose, with the meaning suggested, then how did the driver manage to have that water-line pull loose at *that* time? Would this be an example of psychokinesis . . . of mind over matter? We prefer to think not, but it is an open question.

A Car Lock-Out

On one occasion a traveling salesman found himself locked out of his car. With the keys in the ignition, he had "accidentally" slammed the door in the appropriate way to lock it. The windows were all rolled up, naturally!

What happened 37 hours before?

He had spent the night alone in a hotel room in Colorado Springs, and was awakened shortly after midnight by sounds of sexual violence in the adjoining room. How was he feeling? Like a frightened would-be hero who should break down the door and tell the man to quit it! But he suppressed this urge, without too much difficulty, and turned over with the thought that the female had probably asked for what she was getting.

In order to resolve that whole situation, it would seem that he made sure that no one could enter her (the car)—37 hours later.

The Lost Car Keys

A business acquaintance with whom I was visiting in Seattle spent a futile half-hour frantically searching for his car keys. The car was locked. He couldn't find the keys and had looked every conceivable place. Now it happened that I was present the day before, when he had received a letter from his wife who was visiting her daughter in Alaska. That letter, written from above the Arctic Circle, in the middle of winter, appeared to hold the clue to the mishap.

"Sometimes," I remarked, "if we connect with the trigger incident for something like this, it helps to solve the problem. Remember that letter yesterday—in which your wife said something like 'I feel like I've been here an eternity, but of course this is not your world!' "

No sooner had I made the connection with the letter, than he went straight to a closet where the keys were found in the pocket of a bathrobe. That ecstatic comment in the letter apparently had a much deeper impact on him, and the keys being lost symbolized the feeling of being excluded from her world.

But wait a minute. We have been saying that these "mishaps" are homeostatic; that they serve to "balance the books" in some manner, to resolve subconscious problems. How do we interpret the lost keys? How does it solve a problem? Later on, we will see something of the universal significance of the event of birth, and some of the many ways that this early trauma manifests in a variety of physical ills. But right now, we shall merely affirm that expulsion from the womb, and being cast out of mother's body, has been a problem with which we have all played!

It may be that this frantic search for lost keys provided him with an "explanation" for the exclusion . . . that if he could only find the right key, he could participate in her world, possibly even to returning to mother's womb!

One more comment: whenever we make short simple explanations, please bear in mind that even the simplest of such incidents may be far more complex than we realize, and that we are not claiming to do more than hint at a general area of interpretation.

A Collision

On a Friday morning, about 7, a friend of mine was driving to work in a light drizzle, when a car slowly entered the road from the right, apparently oblivious to his approach. It was a downgrade, and his full pressure on the brakes merely locked his wheels in a free skid straight into the other car's side. At about 35 miles per hour, he was fortunate that his only injury was a goose-egg where his forehead struck the windshield.

Why was this necessary? And what was its meaning?

My friend was on the board of directors of a new enterprise and was in a meeting, about 37 hours before, in which the issue at point was his complaint that the chairman of the board did not trust the group consensus and was manipulating control without adequate discussion of policies and projects. It was a short and sharp confrontation, which ended in his expulsion from the board.

The trip my friend was making (an honest, straightforward trip) was stopped abruptly by someone at cross-purposes to his. And, strangely enough, the driver of the car with which he collided was stone deaf in his right ear, as was the chairman of the board.

When people wish to collide, they collide. And when they

wish to just miss, they just miss, as in the following case, where a clash of wills produced danger signals only.

A Near Collision

Returning from a family camping trip through eastern Washington, I was at the wheel when, after passing through an underpass, I started to make a left-hand turn onto the freeway on-ramp.

It was a few minutes before 10 P.M. as I started to cross the opposite lane, and was not (consciously) aware of a VW bus bearing down on us at high speed. As my passengers shouted, and I jerked the car back into my lane, the other driver swerved violently away from me, and we missed by inches.

Now, the question is often asked: *might* we have had a severe accident if I had not reacted so quickly? And the answer is no, we subconsciously engineered exactly what we wanted, no more, no less, and such speculations on the "what might have beens" is technically meaningless. We can say this because the trigger incident, 37 hours earlier, showed clearly that we *intended* to miss, not hit.

Two mornings earlier, between 8:30 and 9:00, we were breaking camp, and I was helping my brother-in-law secure the sleeping bags, tarpaulins, and blankets on the rack on top of the car. My sister intruded with various suggestions, and there was a clash of wills as to the best way of doing it. Her stubborn persistence won out. Later, she complained that we should grant her credit for at least one little corner of creative know-how after all her years of experience.

In this incident 37 hours later, we dramatized the danger of trying to cut across her path. It was apparently far more important to her self-esteem and very existence than we realized.

Naturally, we're not claiming that *all* car troubles have a precise one-to-one relationship with human psychological problems. But if we find car troubles happening a Psycle or two after a personal problem, and if we can see symbolic meaning to it, should we discount the relationship?

The car just won't get started? Battery dead? Backtrack a Psycle or two and you may well find, sure enough, that there is a deep problem of someone being "dead" about something. Just can't get going. Afraid. Hanging back. Not about to get going.

Even a traffic ticket may be on purpose. One case history happened 37 hours after a lay-off from a job. Feeling insecure, or whatever, he failed to come to a full stop at an obscure intersection, at night, where a cop was just waiting for him. He knew, subconsciously, where to go to get it. And he got exactly what he wanted—a ticket—for whatever need-feeling it meant for him.

A Highway Spin

One three-time divorcée, going into her fifties, was feeling very depressed because she had no real direction in life. Just before a trip north from San Francisco, one of her friends urged her one way, and another friend offered her a completely different alternative, and neither really "grabbed" her.

What happened 37 hours later? Going through a mountain pass, on the downgrade, she hit some black ice, smashed into the right guard rail, bounced back to the center divider, careened off it, and spun around in the middle of the freeway, with cars flashing by on both sides. And that was it. There she was, dramatizing right out of her subconscious the predicament she was feeling about life in general. Just spinning around in the middle of the freeway of life, not knowing which way to go, while everybody else seemed to have their directions and intents and goals.

With knowledge of Psycles, we may have amusing insights into the private lives of public figures, as for instance, even the President.

President Ford's Collision

When Jerry Ford's limousine was struck broadside in the middle of an intersection in Hartford, Connecticut, in October of 1975, was this not just a meaningless "accident"? Although we weren't privy to the exact happenings within his family, we feel certain that either 37 hours before, or 3.1 days before, there was a serious confrontation with his son Jack over his public statements about smoking marijuana. It was in the papers.

We can almost hear the President's subconscious crying,

"Dammit! you're hurting my public image, running at cross-purposes to your dad's best interests. I'd be doing better in my public relations if my own family weren't torpedoing me from the side!" It was a carload of teenagers, that came and hit him from the side, with a young man about Jack's age driving.

Later on, we'll reconsider many cases, to show how they illustrate the existence of a far-ranging subconscious communication—ESP. But for the present, the important point to recognize is that such Psycles calculations and reflections can be immensely useful for our own self-insight and self-liberation. Such incidents can be important clues to what is happening inside, and where to look for solutions.

HOW THE SEXIEST STORIES HANG OUT FOR THE WHOLE WORLD TO SEE!

Given the facts that people are incredibly symbolic, that sex is a universal fact of life, we now come to a category of case histories which is as full of insight for our lives as that of auto mishaps.

The use of Psycles, and letting ourselves *feel* into the symbolism behind our actions, can mean the stark, naked revelation of what used to be some of our most private secrets, hanging out for the whole world to see!

The Near Burglary

When Susie returned to her apartment Monday noon, she surprised a burglar who had collected her most valuable possessions on the bed but fled out the window leaving them behind.

This may seem innocent enough until we add up body-time, Psycles and symbols. From that we can deduce (and it turned out to be true) that she had a tempestuous date Saturday night and almost lost her virginity! At about 11 P.M. the pressure was heavy, but she did not acquiesce, and thereby remained a virgin.

At almost exactly 37 hours later a burglar had gathered her clock radio, stereo, jewelry, portable TV, and tape deck on the bed—ready to be stolen. But it did not happen. It was stopped just in time, and she retained her most valuable possession—virginity.

Does this seem too unlikely?

Just remember that we are dealing with a universal rhythm which is intrinsic to all life from microbes to man. Called the circadian rhythm, or "biological clock," it shows us for the first time the connection between many disturbing stimuli and their delayed reponses—a causal relationship we didn't realize existed.

In this section we will scan a number of cases which are sexual in nature, to show how the Master-key works. Most simply, all we have to do is remember what happened 37 hours *before* the trouble. It almost always works. The variations (other multiples of 12.5 hours) and problems we'll discuss later.

The Honeymooners

George was a junior engineer in aerospace, who had returned to his office on Monday after a weekend honeymoon following a Saturday wedding. At 10:30 that morning, while walking around a desk, he "accidentally" caught his fly on the corner of the desk, and ripped his pants wide open.

Was this just an "accident," and meaningless?

Looking back 37 hours before, we didn't have to embarrass George by asking what happened Saturday evening about 9:00 in their honeymoon cottage at Lake Arrowhead. For us it was a haunting reminder that many newlyweds have troubles on their wedding night, especially if they are young and innocent.

The Lathe Bolt

Elsewhere in the same aerospace plant was Henry, a machinist, working with a lathe, and machining down a large plastic tube. He overtightened a lathe-bolt, snapped its head off, sprang the gripper, and the heavy tube fell crashing to the floor.

Was this just "one of those things"? Meaningless?

Thirty-seven hours earlier, Henry was visiting with his future father-in-law. He was asked about the visit to his fiancée's home and whether he had any troubles with her father.

"Hell, no," he exclaimed. "Just having a couple of beers with her old man . . . no troubles . . . just boozing it up, but that s.o.b. sure has some screwy ideas about sex!"

Apparently the father was laying down rules of sexual conduct for Henry to follow in his relationship with the daughter. One large phallic symbol lay on the floor in mute testimony to the conflict, and its aggressive resolution.

The Hot Pot

In a different chemical engineering plant, John was working in metallurgical research and found himself, one morning, overseeing a huge stainless steel cauldron filled with molten titanium.

Somehow, the electric controls stuck (or were manipulated wrongly) and the pot was overheating out of control. In a near panic, he had to pull the master-switch, cutting power to the whole building.

Was this a meaningless "accident"?

Hardly! Two evenings before this incident, he was scheduled for a date with a very sophisticated woman. With deep conflicts between a homosexual problem and "proving" his manhood, he canceled out at the last minute. In his subconscious mind he apparently "balanced the books" 37 hours later, with a "proof" that he did the right thing. The date would have been much too hot, and very dangerous!

Boswell in London

Some critics have argued that the 37-hour timing of Psycles may be biased by suggestion, and be an example of what is called "self-fulfilling prophecy." According to this theory, the subject may unconsciously bring about some mishap at the 37-hour time because it is expected of him. If this were so, the entire concept of Psycles would be false, and based on positive suggestion, or strictly coincidental.

We therefore went back a hundred years, long before anyone ever thought of 37-hour time-lags between trauma and symptoms, and studied the detailed diaries of James Boswell who was famous for his *Life of Johnson*. In his London Journal, he described his day-to-day life with unusual detail and candor. We found a number of good Psycles cases.

One of his main problems in London was the effort to get a

genteel job with the Royal Guards. After many failures he confronted the last and final "power" who could swing such an appointment. It was Lady Northumberland, and she turned him down.

Approximately 37 hours later, judging from his diary, he entered a pub, rang for the manager, and requested the biggest and fattest whore in the district. When this creature entered, Boswell proceeded to haggle over price, and ended up screaming at her for demanding too high a price.

"Get that monstrous, greedy bitch out of here!" was how Boswell settled his accounts with Lady Northumberland—in his head! One wonders whether this good lady made certain sexual demands on Boswell which he considered too high a price for accepting the appointment.

The Signpost

While visiting a social club recently in the northeastern sector of Seattle, I was posting on their bulletin board one of my dial calculators. I had lectured there about a month previously, and had had other discussions with some of the members.

Two men approached.

"I have a case history for you," said one of them. "When I heard you speak last month, I didn't buy your idea at all, but now I'm a believer. See this scar on my forehead?" There was a faint red line about an inch and a half, running diagonally on the upper right quadrant of his forehead.

"Well, about a week after your talk, I was jogging down by the lake, and ran smack into a signpost while cutting across a corner. And you know? Exactly 37 and a half hours earlier, I was in bed with my girlfriend, and was being told off, that this was the last time, that she was through with me, and that our romance was over."

"What did the sign say?" I asked.

"All traffic must turn right." A quizzical smile crossed his face. "Actually, there were a series of signs of our impending termination, which I didn't notice, or pay heed to . . . like, I should have known it wasn't going to last much longer."

I suspect that if we had taken a deeper probe of his life,

personality, and problem themes, we would have discovered a chain of significant traumas and events having to do with signs, and about paying attention to them.

The Thumb Gouge

The other man was a newcomer to the whole subject. He happened to have a fresh wound on his thumb.

"What's this all about?" he asked, pointing to a calculator I'd devised to quickly figure out Psycles time-lags.*

"With this calculator, we can often pinpoint the hidden causes for accidents."

"Like this thumb? Go ahead and show me."

"All right. The first question is: just what did you do to your thumb, and the second question is—exactly when did it happen?"

"Happened this morning, about 9. I was using a claw hammer opening some boxes, and slipped, and the sharp claws of the hammer went into my thumb. Kind of embarrasses me, because I am usually very careful, and as a teacher, I show my students how not to get hurt."

"All right. Now we set the 'mishap' pointer at 9 A.M. Wednesday, and you see here . . . this indicator. What were you doing Monday evening about 8? Thirty-seven hours earlier."

For over a minute he couldn't remember what happened Monday evening. But then it broke through. "Now I remember. Met a gal in the park during the weekend. Hadn't seen her for over a year. Monday evening I went over to her apartment, hoping for some quick and ready excitement. Like she might be available. But she told me she was busy and didn't want to see me."

People don't usually like to admit failures, or setbacks to their egos, and some of the deeper feelings involved.

"I think your subconscious is trying to tell you something," I ventured rather carefully. "If we look at exactly what you did about it, being turned down, and realize how symbolic we are, it might be as though that kitten has sharp claws and could hurt you. I like a warning—like it should be 'thumbs down' on her, stay clear, she's dangerous."

*See page 259.

"Hmmmm . . ." as though faint flickers of insight were breaking through. "Yeah . . . you could say that."

Within the context of our cultural inhibitions in the area of sex, there is, of course, a wide spectrum of behavior related to problems of guilt. What do we do, and how do we respond to our inner feelings, when we violate religious teachings and social prohibitions?

The Hot Hand

In our society and most societies, sexual feelings within the family, directed toward each other, are of course totally disallowed. They exist, most naturally, between fathers and daughters and between mothers and sons, but they are heavily repressed for the most part, and appear only as symbolic outbursts. In the following case the feelings which erupted momentarily were followed by a form of self-punishment.

A female laboratory technician picked up a metallic rack, not consciously realizing that it had just recently been removed from a furnace and was still very hot. She burned the fingers of her right hand.

Inasmuch as the time was close to 5 P.M., it seemed rather pointless to ask her what she was doing about 3:45 the previous morning, although sometimes we do find dreams, nightmares, or other nocturnal happenings.

"My son," she replied, "was leaving for military service overseas, and I went with him to the airport, and, right then, was walking hand in hand with him to the gate. Was the same hand . . ."

She was a widow, and this was her only son, and from the injury I would suspect that there might well have been some deep feelings—sexual—which surfaced at that moment, but were immediately suppressed, and for which she felt guilty.

Hence, the self-punishment, 37 hours later.

The Open Camper

An artist-wife discovered with great consternation one morning that she had left the VW camper with its tailgate open all night in the alley behind their house. Inside the camper were some $2,000 worth of paintings, untouched.

The sly husband, thinking back 37 hours, instantly knew approximately what happened. Because at that time, she had taken some paintings to a doctor's home for a private showing. Judging from the camper incident (symbolizing her body), he suspected that she had presented not only her paintings, but an offer of bodily delights as well, wide open for the doctor to take and enjoy—but he had declined the invitation. Her intended infidelity had been rejected. Technically, of course, this case history is flawed by the fact that it rests on pure assumptions. (The husband never had the courage to ask his wife what really happened. They were later divorced.) We include it, however, because the symbolism is typical and instructive.

The Bike Detour

A university coed in her junior year, cycling down a steep grade, came within inches of a nasty collision when a car made a sharp right turn in front of her, into a park. With split-second braking, she managed to swing alongside this car, making an inside turn, also into the park entrance. Could this complex event be purposeful, and with some deep symbolic meaning?

About 37 hours earlier, she was experiencing a crisis of sexual identity, and canceled a date with a young man to go out with a female friend who was, with considerable power, enticing her into a lesbian relationship. The car symbolized the powerful other woman changing the course of her life, if we are to accept the evidence of the circadian timing as proof of intent. Here again, noting the perfect cooperation involved, we have the implication of subconscious communication, such as ESP, which we'll explore in more detail in Chapter 12.

Why Probe the Hidden Reasons?

Although the foregoing stories may be amusing in some respects, we should realize that each accident or mishap may represent hidden and unresolved pain. In each case there is a person with a secret problem, conflict, or tension. If we have such troubles without gaining deeper insight into their hidden cause, we are missing a unique opportunity to progress in our search for freedom.

An accident can merely "balance the books" subconsciously, without contributing materially toward our advancement in self-liberation. But if we pounce on it, while it's still wriggling, and find out why we did it—then we propel ourselves one giant step toward that ultimate freedom. We can become that much less conditional, unconscious, "hung up," bound, and blind. Each accident or mishap, analyzed, releases a portion of our vast and truly divine potential.

3 A New Understanding of Disease and Healing

Part A. The General Problem and the Roots of Confusion

BEFORE we enter the whole area of disease and show what a contribution Psycles can make toward the understanding and healing of many internal troubles, it would be best to review a number of questions and assumptions about disease and healing in general. Compared with "accidents," which are relatively simple in theory, this area contains many complex and contradictory beliefs.

In Part A we will first show something of how and why it is that we deliberately avoid coming to grips with the problem, and are content to wallow among scores of conflicting ideas. In Part B we will offer a road out of this quagmire by introducing a simple and comprehensive new theory of disease and healing which derives inexorably from our revolutionary discoveries in the 2,000 Psycles case histories.

Techniques and Theories without End

For us, as living organisms, there are a thousand ways and places where things can "go wrong." Each structure or function of the body-mind has its list of distortions or "diseases." And throughout history enormous efforts have been made, ostensibly, in trying to discover why they happen and how to end them.

At the present time, science does not really know how an aspirin turns off a headache, nor how passions turn off painful symptoms, nor how tens of thousands of drugs work, nor how nutrients or physiotherapies can modify behavior, nor why sometimes people just get well when they should die.

And yet in the world today, for treating and/or "curing" these myriad troubles, there are hundreds of different healing systems, using thousands of different healing techniques. And behind each of them are staunch adherents who can vouch that in some manner, at some times, they have all effectively *worked*. Associated with them, also, are hundreds of different theories and explanations that contradict each other in bewildering fashion about *how* or *why* they work.

Some of this diversity is illustrated on page 41.

Judy's Throat Hemorrhage, as Reference Point

In our introduction, as one of the three case history examples, we told briefly the story of Judy and her throat hemorrhage. When we touched on her confrontation with her mother, the hemorrhaging stopped immediately. The connecting of the "cause" and the "effect" was in this case therapeutic. There was sudden insight, and sudden healing. But Judy's throat problem could have been "cured" by an antibiotic, or by a faith-healer, or a chiropractor, or a hundred other systems of healing.

Using Judy as an example, let's explore this fact.

A Look at Nine Different Healing Systems

If Judy had first gone to a spiritual healer, she could very well have been healed following some intensive prayer in front of a large audience and a sudden laying-on of hands. Her cure would be explained in terms of sin, repentance, and divine intervention.

Or, secondly, she might have gone to a psychotherapist. Here she would experience a lengthy dialogue, involving free association, dream analysis, and the pulling together of some "good reasons" to provide "insight" into unconscious causes. Her cure

A. Among the hundreds of systems which promise a better life:

Shamanism	Monasticism	Herbology	Health spas
Ayurvedic	TM, TA	Acupuncture	Yoga
Medicine	Synanon	Astrology	Mind dynamics
Chiropractic	Living Love	Shiatsu massage	Voodoo
Homeopathy	Subud	Psychoanalysis	Spiritualism
Christian Science	Zen, Tai Chi	Sufism	Theosophy
Kabala	Arica, est		Aikido

B. There are thousands of methods and techniques of treating or poking the body-mind:

electrolysis	hot/mud baths	carbon dioxide	dream analysis
massage	prayer	fasting	radionics
medicinal drugs	meditation	color therapy	concentration
mineral water	mantras	positive thinking	psychodrama
alkaline salts	relaxation	chanting	nude therapy
diets	autosuggestion	travel	encounter
novocaine	yoga asanas	sun/steam baths	vibrators
megavitamins	yoga breathing	marijuana	biofeedback
mustard plasters	behavior modification	exercises	grape cure

C. At or in such places as:

stomach cavity	eyes	joints	lung cavity	intravenous
spinal column	nose	muscles	skin surface	intramuscular
ears	mouth	nerves	blood	rectal

D. So as to counteract and help turn off such troubles as:

cancer	stomach aches	crimes	burns
heart attacks	backaches	hypertension	hives
migraines	hay fever	colitis	colds
asthma	depression	injuries	losses
hepatitis	acute anxiety	mental aberrations	tachicardia
epilepsy	anemia	hypoglycemia	diphtheria
ulcers	cataracts	nephritis	mumps
obesity	liver troubles	malaria	vertigo
arthritis		rheumatism	

E. And there's no end to the many different theories and explanations for how they work.

would be explained in terms of conflict, integration, and adjustment to reality.

Thirdly, Judy might have gone to a physician with medical education, and been subjected to a variety of examinations and tests, followed by an antibiotic injection. Her cure would have been explained as an outcome of the chemical destruction of an infectious agent.

As a fourth option, Judy might have been treated by a chiropractor with spinal manipulations, heat, and electrolysis, and her cure could be explained as an adjustment which restored nerve energy flow to a spinal nerve subluxation.

Fifthly, with the same throat hemorrhage, Judy might have been successfully treated by an acupuncturist by means of the manipulation of several needles in her skin. This would be explained as the restoration of chi energy balance, by countering the blocks in the body meridians.

Or, Judy might have taken a sixth option by seeking treatment by a voodoo priest, and following ritual chants, incantations, drums, amulets, and potions, been quite cured. The priest would explain that he had driven the evil spirits out of her body.

As a seventh option, Judy could have been treated by a homeopathic physician who had an array of medicinal potions of extreme dilution. With the theory that like cures like, he would have opposed her symptoms with agents which induced the same symptoms in normal persons, and she would have been cured.

Eighthly, Judy might seek the aid of an herbalist who, after supplying her with special herb teas, might explain that he was supplying her with some missing nutrients, which should counteract her blood problem.

And finally, as a ninth option, Judy could have been cured by a naturopath who prescribed a 30-day water fast followed by a diet of grapes. This would have been explained as a method of eliminating accumulated toxins and poisons underlying her diseased state.

Each and every one of them could have cured her, despite all their different explanations. These nine completely different treatments, as well as hundreds of others, could all have healed the same identical and real physical trouble.

Beyond these nine systems, there are hundreds of other

explanations for why people get sick: planetary influences; the lack of certain colors of the sunlight; bad vibrations; past life retributions; hexes; hereditary predispositions; genetic weaknesses; love of suffering; seeking of pity or attention; the weather too hot, cold, wet, or dry; the body system being too alkaline or acid; the metabolism being "out of balance"; cells, tissues, and organs just "going wrong"; organs not functioning properly or wearing out; not eating the right foods or getting the wrong food combinations; not getting the right mineral elements, or getting them in the wrong combination; and so on.

Behind each of these systems and theories, there are vast numbers of people who swear by them. How, therefore, do we deal with the fact that there are so many different methods of treatment, each with its believers? Are most other people deluded? And how do we relate to the hundreds of different explanations for all these therapies? Are they all wrong except ours? How do we react to a method which obviously works but whose explanations seem to us bizarre? If we went to China and saw acupuncture work in an open-heart surgery, would we come back and shake our heads and say no, it can't be valid because it doesn't fit our theories? Or, on the other hand, because a method works, does that necessarily mean that its theories on the causes of disease and health are correct?

Question: With our revolutionary new data, is it possible we can cut through all the diverse and contradictory explanations of why we get sick and what makes us heal and find an answer to the whole picture, one that will be "true" a hundred years from now?

It's not enough to come to the simplistic conclusion that it's "all in your head." We need to be able to understand—physically—how it can be that each of these scores of totally different healing techniques can be successful in treating the same disease, when at the same time each of the healers believes something totally different from all the others!

Before we show what happened in Judy's case, and the mechanics of healing, and how they all work, let's first consider the existing situation and ask how it is that all this diversity and confusion could have existed for so many thousands of years.

The first part of the answer comes from the simple fact that we are totally psychosomatic, but make strenuous efforts to deny it.

Of Course We're Psychosomatic—But—

In the early 1940s the word "psychosomatic" came into popular use with the realization that we did, somehow, have something to do with our troubles. Most of our bodies had heads. Hence, there were minds and hearts involved in the genesis of our problems. Our mental functions (psyche) and body structure (soma) were intimately involved with each other as parts of one whole. It became fashionable for medical authorities to declare that from 70 to 90 percent of all patients seeking medical care were suffering from psychosomatic ills which have emotional roots as their primary cause.

There are two reasons for this:

Firstly, we are composed of trillions of cells, each of which is a beehive of electrochemical activity, and there is nothing we think, feel, or do, which does not manifest as physical changes within. Having heads and hearts, we are mind-body systems.

Secondly, it is a fundamental fact that human problems such as isolation, failures, insecurities, frustrations, and insoluble conflicts, with their emotions of anger, fear, hate, grief, or anxiety—when intense and prolonged—all cause adverse effects throughout the body. These range from aches and pains, through metabolic and cellular distortions and a wide spectrum of diseases, all the way to death.

What we think and feel governs the whole body. And that means that we are all absolutely psychosomatic. The physical basis for this was spelled out by the great names in the field—Walter B. Cannon, Flanders Dunbar, George W. Crile—fifty years ago.

The large body of evidence from faith-healing serves to confirm this. Dr. Alexis Carrel wrote an immensely popular book at that time *(Man the Unknown)* calling attention to the healings at Lourdes in France. These "miraculous" cures involved many gross and incurable distortions of the body. And yet, without physical treatment, the conditions reversed for rapid and complete healing. Did this not prove we were psychosomatic in nature?

Evidence from hypnosis shows that we control practically every cell of our bodies. Through mental commands alone, we can modify our tissues, cause them to inflame, proliferate, shrivel up, overreact or become erratic. Again, psychosomatic.

Why We're All Psychosomatic: The Physiology of Feelings/Emotions

In nature—when our lives are threatened—the whole body-mind is galvanized for survival! Thus:	But now, the survival threat of financial insecurity, social rejection, isolation, fear find no outlet in violent action, thus . . .
The HEART is speeded up, circulation speeded up, metabolism increased, temperature elevated.	Strain on heart, myocarditis, heart degeneration. Maximum production of waste products. Strain on kidneys. Cardiovascular diseases, arteriosclerosis.
The BLOOD is transferred from stomach, intestines, and genitals, to heart, lungs, central nervous system, skeletal muscles for supreme fight or flight action.	Rise in blood pH, increased acidosis. Inhibition of digestive functions, secretions, intestinal peristalsis, putrefaction, indigestion, autointoxication; more strain on organs of elimination; saliva changes, tooth decay. Blood coagulation time reduced.
ENDOCRINE secretions including adrenalin and insulin for super-energy; thyroid stimulated.	Emotions intensified; skin moist and soft; eyes staring, protruding. Excessive glycogen released from liver.
NEUROMUSCULAR activity stimulated, contractions of smooth muscles of skin, hairs stand up.	Limbs trembling, heart pounding, respiration rapid, mounting temperature and metabolism. Extremities tingle; hypersensitivity, neurasthenia, less organic activity.
Special SENSES rendered more acute; pupils dilated, masking of pain sensations.	Raised blood pressure, excessive perspiration; impaired circulation.

Lacking *action* to "burn up" the chemical changes, the organism suffers.

Laughter and *crying* are explosive types of action which help to discharge the tension/distortions.

Fear, worry, disappointed love, anger, hate, rage, anxiety—are all non-action types of responses which, with their chronic physiological distortions, are organically destructive.

Unfortunately, there is still much suspicion on the part of the general public and even the medical profession, which favors "physical" rather than mental-emotional causes. It's almost as if calling something "psychosomatic" is like saying a person is crazy. Which, of course, is not true.

To really face up to our psychosomatic nature means a very difficult assumption of self-responsibility for some shocking events, and a willingness to confront some sticky motivations underlying the painful things we do to ourselves. Deep within all of us is the knowledge of self-responsibility and the fact that painful truths underlie our suffering.

To avoid this truth—which we fear almost more than death—we expend a great deal of energy in the effort to prove that we are not psychosomatic, nor responsible. We prefer to think of ourselves as innocent victims of chance forces "out there," acting on us against our wills and without our consent or desire.

This is reflected in verbal dodges, in our preferences for treatment, in our language, and in our acceptance of pseudo-explanations.

Our Verbal Dodges

Until Psycles, one of the biggest problems in the assignment of causation was the vagueness of the psychosomatic finger-pointing. It was extremely difficult to find the hidden motivations. Freud and Jung and many others spent their entire lives searching unsuccessfully for that magic entry into the subconscious. Now and then a case could be tracked down to its psychological genesis, but resistance and blockades to such revelations were mountainous, and the results were rarely sharp and clear. Complex theories evolved, with deep differences of opinion among the authorities. We could therefore ignore the few glaring examples and continue on our innocent way.

A convenient dodge lay in the cure of many so-called psychosomatic troubles treated by physical means such as nutrients, minerals, vitamins, or drugs. Can a physical means be used to heal psychological troubles or psychosomatic ills? And if so—does this mean that maybe they weren't psychological in the first place? That looked like a good escape clause. Also, we were confronted with an uneasy feeling that being psychosomatic meant being crazy—as though something were "wrong" with our minds. And who wants to admit being crazy?

Oh, yes, we were responsible, but it was because we were not being careful, or weren't watching out properly, or paying sufficient attention, and so on. These verbal dodges allowed us to continue feeling innocent and non-responsible. Or, we could avoid the truth by saying that psychosomatic meant that "emotions were a factor." This is clearly a dodge, inasmuch as if emotions are only a factor, then the real cause must be something else "out there" still!

Look at Our Paranoid Language!

If we take a more searching look at the language we use from day to day, it would seem as if the entire cosmos were conspiring to subject us, as innocent victims, to a host of indignities. We seem to be attacked, ravaged, and buffeted by a wide assortment of outside causes.

Notice the words we use: somehow our defenses break down, we fall prey to the common cold, are driven to drink, crippled by arthritis, struck down by polio, succumb to hepatitis, are beset with allergies, plagued by stomach ulcers, stricken by tuberculosis, invaded by cancer, and finally attacked by our hearts!

The language implies that we are innocent victims, in a linear type of thinking, in which forces "out there" are doing things to us against our will. Nowhere is there any hint that we might be responsibly involved; we're always on the receiving end.

Notice also, the semantics of how we describe ourselves as laboring under stress, being paralyzed by fear, burdened by guilt, poisoned by hate, overcome by anxiety, or totally confused by our emotions. It's not as if we generate these emotions for reasons. No, *they* do us in. It is as if our organs, or even vague and complex emotional functions which don't even belong to us, are our enemies, endowed with separate wills, waiting with evil intent to pounce upon us the moment we let down our guard. In this essentially paranoid view, we say we are vulnerable to stress and prone to accidents, as if they were things acting against us, rather than something we do.

This attitude of being an innocent victim, irresponsible and uninvolved, permeates our language, and hence also our thinking in connection with the whole problem of human suffering. In this, our greatest illusion, troubles happen *to* us, against our will, and without our desire or consent.

A Thousand False Explanations

The honest healer would say that he does what he does, and it often works, but he doesn't really know why. The point is that practically all healing methods, drugs, and manipulations have their origins in trial and error. For some particular suffering, something was tried, found effective, and tried again. This is empirical knowledge, accumulating equally for the cavemen as for the modern scientist, and the subsequent add-on explanations in either case have about as much validity—especially when the scientist caters to our avoidance of internal truths.

Take the lowly aspirin. Between the time it enters the stomach to the time it turns off a headache, there is little knowledge but scores of theories and contradictions to cover our ignorance of how it works. Despite the thousands of marvelous drugs which may counteract almost every conceivable malfunction, there remains little true knowledge of how any drugs work and little understanding of true causes in medicine and the rest of the spectrum of healing.

In avoiding admitting that we're psychosomatic, we go to healers and say, "Please turn my pain off—but don't touch *me*!" Whereupon we undergo an incredible variety of ritual with its attendant mumbo-jumbo, experience relief, and think well! his theory must be true because his method worked! Which is nonsense, of course, because although a method may work that is no proof the explanation behind it is correct.

Now, despite our desire for safe, external explanations for why we suffer, there have been a goodly number of substantial ones around for a long time—such as chemical poisons, carcinogens, nutritional deficiencies, infectious agents, microbes, or heredity, for example. We will see how even these explanations become suspect, with Psycles.

On Ropes and Ladders as "Causes"

To help illustrate a common problem in the assignment of cause, we will go to the analogy of a prison break. In the accompanying sketch we see a portion of a prison, with a prisoner in the window, a knotted rope hanging from the window, and a ladder leaning against the prison wall.

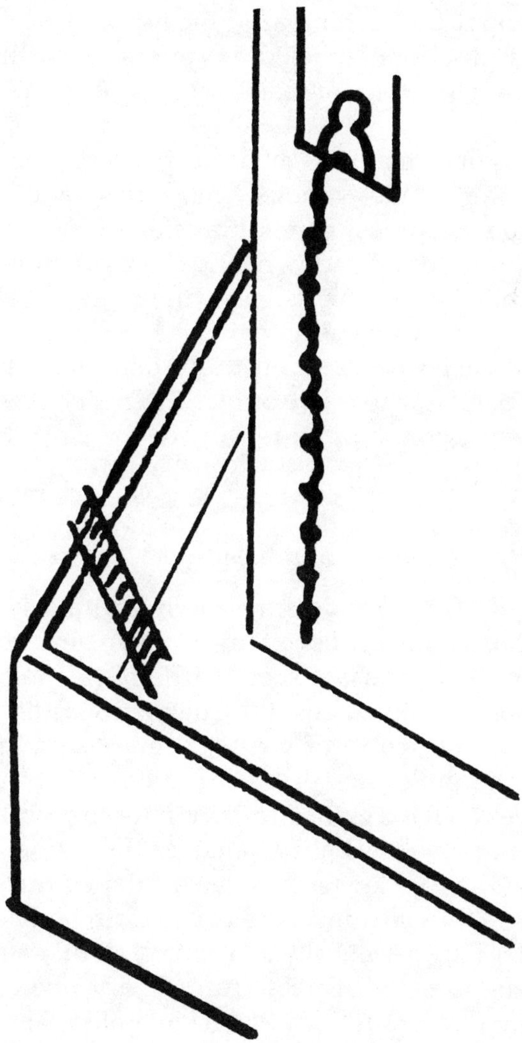

The problem is to analyze the cause of the prison break, to answer why it happened, and to point toward its solution: the prevention of future breaks.

One obvious answer is the knotted rope; if there were no rope, there would have been no prison break; therefore the rope is the cause. Or, one could single out the ladder with the same argument: if there were no ladder, the break would have been impossible; therefore the ladder is the cause.

Throughout the healing arts, including medicines, we often find ropes and ladders labeled as causes, with no acknowledgment that there is a human being, up there in the window, who wants out!

In the same manner medical researchers, searching for "causes," single out various molecules, viruses, enzymes, metabolic abnormalities, genes, transfer RNA's, nutritional bits, vitamins, hormones up or down, and scores of other variables, and say—Ha! There it is! The cause—when they are all ropes and ladders, or links in a causal chain.

Is it not easier to settle on these links than to consider the patient as a person who is in trouble? In the diagnostic nit-picking, they even settle for Rung No. 3 or maybe Knot No. 6, and say There! That's the cause!

And Now—Molecular Biology!

From out of thousands of research laboratories exploring the secrets of life came a host of marvelous new reasons for our troubles. Tens of thousands of biochemists, probing with new instrumental breakthroughs the microstructures of cells, were laying bare a fine world of viruses, genes, and enzymes.

It was the molecules that were doing us in!

Now we can have great admiration for the work being done in these laboratories. The ability to analyze the amino-acid sequences of the six individual legs of a virus and spell out approximately where on the DNA chromosome in the main body of the virus the individual legs are genetically controlled, is truly amazing. Realize that the virus is quite invisible to ordinary microscopes.

With this enormous body of data in molecular biology, with its endless chains of metabolic reactions, the scientific establishment found a very large number of *things* to which it can point as "causes" for our troubles. An excess of this molecule or a deficiency of that one, viewed from their bias, can appear as a neat and logical reason for a "disease." In fact, the data is so massive that psychologists and mystics who would argue for the notion of self-responsibility might easily be stampeded into throwing in the towel, to quit their hopeless struggle, and go along with the tide of innocence.

But hold on! Before throwing in the towel, let it be stated that there are other possible explanations and interpretations of the facts. No question about the factual data itself—just the matter of how it's interpreted. For instance, something may be considered a "cause" when it's actually a resultant in a distorted state of activation.

Part B. The Solution: A New and Comprehensive Understanding

We have been dragging our collective heels for a long time, avoiding the sticky answers to the mysteries of disease and healing. Having seen some of the reasons for this in Part A, let's proceed beyond these roadblocks toward a more final solution.

As you can easily see, the revolutionary discoveries of Psycles go far and wide, inviting us to give up many old beliefs about how and why bad things happen. Since we don't readily give up old beliefs for a new reality unless it *feels* both logical and understandable, we therefore need a new model which fits with our experience, is acceptable with what we know for sure, and on which the new data can hang together and make sense.

Here, then, is our new "bedrock" model for understanding Psycles as it unfolds in the next chapters. First, we have to introduce a new way of viewing our basic structure and function, answering the fundamental questions of how we're built and how it all works. We may *seem* very complicated and difficult to understand, but we're changing all that. This concept is quite simple, yet backed by a great deal *of* technical evidence, some of which can be found in Appendix G for those who may wish to explore the roots in greater detail.

From this new vision of how our structure works, we'll then show how most, if not all, diseases are generated. We will illustrate it with our classic case of Judy and her throat hemorrhage. We will then cap this with the "final" solution for understanding all healing.

A Simplified Model of our Structure: How We're Built

What we *are* is a 3-D network of cell-structure, like an oscillating flexible crystal. We're really quite simple and comprehensible. We have bi-lateral, mirror-image symmetries which make us quite orderly and neatly put together.

Let's see what this means.

What is a *crystal*? In the simplest example, table salt is composed of sodium and chlorine atoms, alternating in three directions: Na, Cl, Na, Cl, Na, etc. (See Figure 1.)

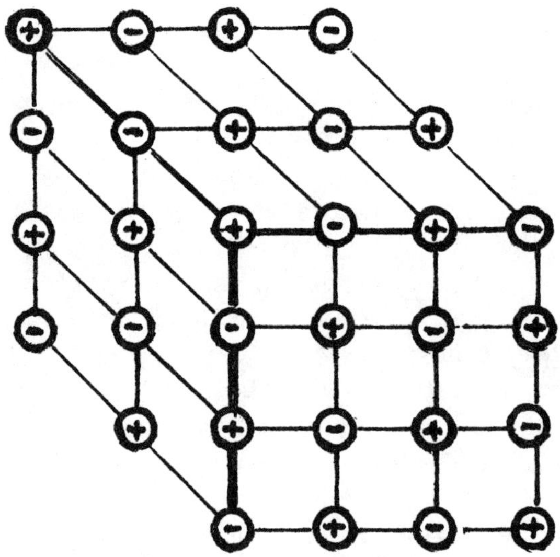

Fig. 1

In the body, it's cells instead of atoms, alternating in polarity (plus-minus-plus-minus) in three directions also. Because we all started out as a single cell which divided and divided and divided and ended up as trillions of cells all hanging together as one entity—there is an order which transcends what *seems* to be very complex.

Not only are we a comparatively simple 3-D crystal network, but we are also hardly more than *six cells deep* from surface to center. All our tissue cells are connected to the brain (cortical centers) by nerves, as pictured in Figure 2. And as the diagram shows, a pin jab in a muscle is only two or three nerve cells distant from the brain, and another three cells distant from the stomach.

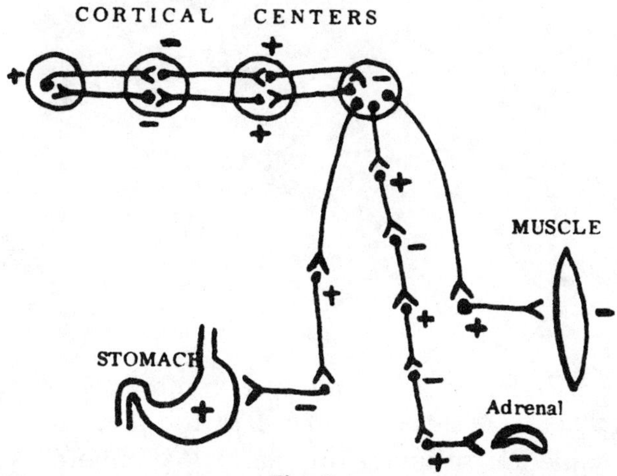

Fig. 2

Thus, throughout the body, our trillions of cells are all marvelously and closely interconnected, with just a few cells (nerve connections) separating any two cells anywhere. This is shown in its entirety in a revolutionary picture (see Appendix G) in which all the nerve-cell connections (circuits) in six tomes of neuro-anatomy were integrated into one master diagram. We need this kind of information in order to be "holistic" in our understanding of life.

One reason we *seem* so complex is because our crystal structure is *flexible*—doing a lot of stretching here and there, but still keeping its connections, as in Figure 3. A microscopic nerve cell can stretch out two feet or longer! And when cut, it can grow out to re-connect at considerable distances.

The next thing you need to notice is that mirror-image symmetries are found throughout our structure. At the micro-level, this occurs naturally through cell division (which is how new cells are generated). Before cells divide, the molecules that make up their structure must first duplicate and then separate into two

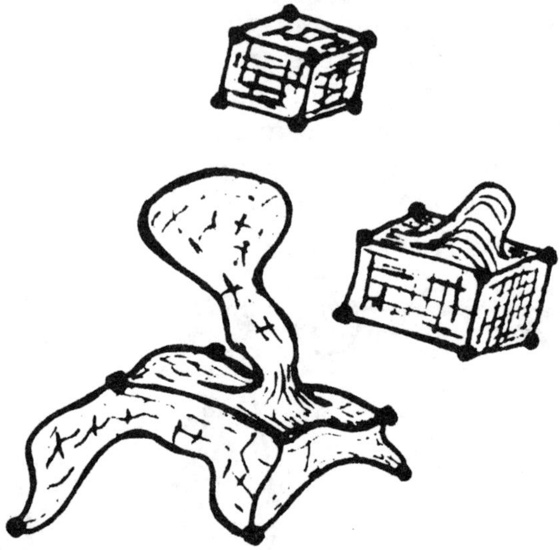

Fig. 3

equal halves. In Figure 4 is a molecule of immunoglobin, one of the many mirror-image structures (A–B is the fold-line.)

This same mirror-image (or bilateral) symmetry is easily seen for the healthy body viewed from the front or back. The two body

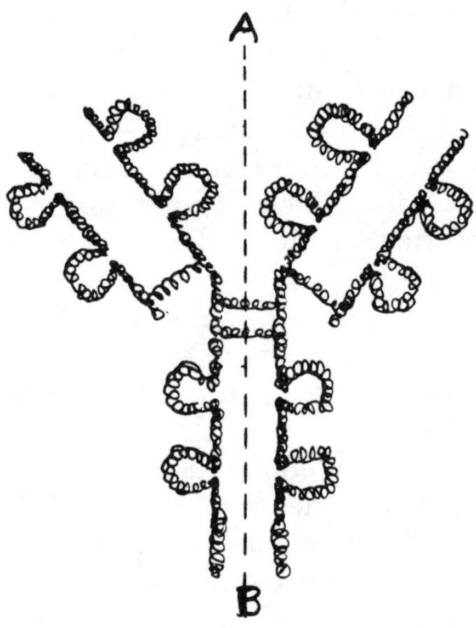

Fig. 4

halves are equal and balanced, whether it be the bony skeleton, the brain and spinal nerves extending through the body, the arms and legs, the sense organs such as eyes, ears, nose, and tongue, or the whole musculature of the body. All are balanced through the center line.

But did you realize that the internal organs, which look quite jumbled and complicated, are also symmetrical? Through a number of steps, the details of which can be found in Appendix G, we arrived at the two pictures in Figure 5.

Fig. 5

The left-hand picture shows the total neuro-anatomy, with the central needle being the brain and spinal cord, with arm and leg extensions. Spraying out from top, center, and bottom are three elements of the autonomic nervous systems, with the sympathetic ganglia lying along the two sides of the central spinal cord. All of these nerves connect with all the tissues and organs of the body—shown cut in half on the left- and right-hand sides. Ignore the details (which will be found in Appendix G). All we want you to do is see the *pattern*. Now note the *pattern* of the right-hand picture. It depicts the whole blood circulation. For those with elementary knowledge of physics and electromagnetics, the left-hand pattern is "electric" and the right-hand pattern is "magnetic" in configuration.

We are sure that these electric and magnetic patterns are not just happenstance, inasmuch as symmetries go hand in hand with electromagnetic events. Although many details remain to be worked out, there is a continually growing awareness that electric and magnetic forces are guiding factors in life processes, down to the finest molecular events.

What we are doing is guiding you, through these clues, toward a totally revolutionary understanding of life and away from the "chemical" point of view—toward the basic physics of life. Modern biochemistry, with its random and statistical chance way of thinking, has been leading us on a wild-goose chase long enough. The life sciences are due for a total upheaval. (See Appendix H.)

As flexible crystals, we are also *oscillating*. Living cells are not quiet; their molecules are continually vibrating, doing a heat dance

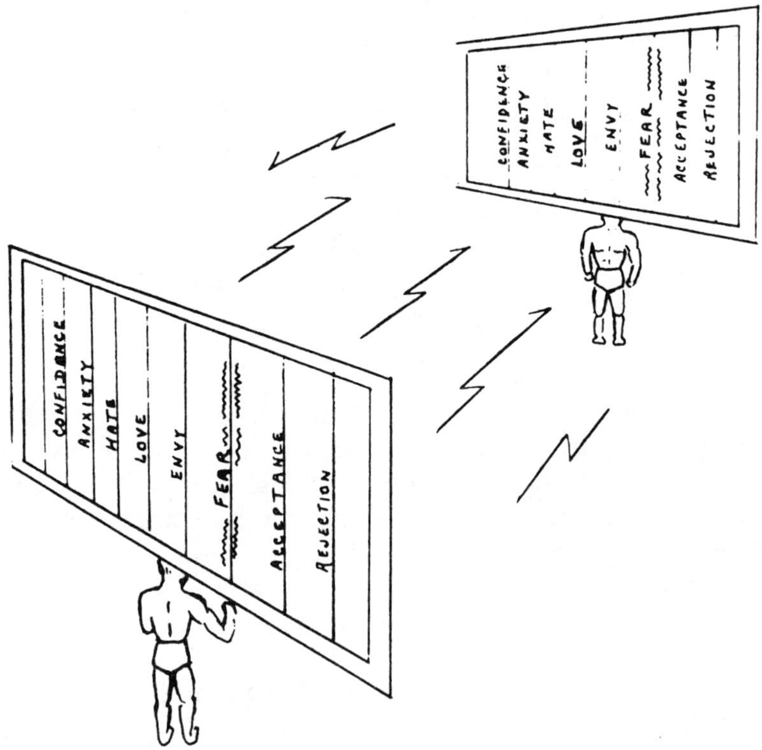

Fig. 6

(thermal agitation). All molecules also have their own natural built-in frequencies. And, just as a piano string will vibrate in tune with a tuning fork that has the same built-in frequency, so the molecules of life are "resonant" (in sympathetic vibration) worldwide.

More complex states of activation, such as thoughts, feelings and emotions, also have their beats and harmonics, as illustrated in Figure 6, and all this leads naturally, according to the laws of physics, to a vast, subconscious, global communication, called ESP, which we will discuss in a later chapter.

A Simplified Model of Our Function: How We Work

How our structure works depends on what *self-image* we adopted, and how we *maintain it* (our homeostasis) in a 3-D cell network in which activations are reciprocal, cell to cell.

Again, let's look at the elements of this statement to see what they mean.

The *self-image.* How we identify ourselves is crucial to our entire life. All infants experience survival threats, or events which are interpreted as a threat to life. The trauma of birth is typical; parental anger and rejection are also seen as life-threatening.

At an early age, sometimes intrauterine, usually by the age of two, sometimes triggered by the first contact with death, we all decided to identify in some manner which was associated with maximizing survival. We adopted an identity, a self-image, which was being *like* this or that model, person or pattern, and acquired a "personality."

The whole process of identifying is discussed in Chapter 15. Briefly, what happens is that babies do a lot of experimenting with resonance, tuning into various signal-structures, acting roles, reflecting, like little mirrors, the various behavior patterns in their vicinity, mimicking feelings, expressions, etc. The reason for this is that the state of resonance (being in tune, in sympathetic vibration) magnifies one's beat, amplifies one's sense of self-ness, and therefore is intrinsically pleasurable and life-enhancing.

But tuning repeatedly into mother's subconscious and painful signal patterns can cause her to react with a barely controlled murderous fury, and be cause for baby, afraid for its survival, to mirror a father-signal for example which may bring responses of love. The decision to identify, therefore, follows from both negative (survival threats associated with what not to be) and positive (love-reinforcement for what is acceptable) signals.

As this identification (self-image) is reinforced by a series of survival successes, (i.e., it "works"), it gets "locked in" (see

Appendix G) and becomes so automatic that it requires little or no effort. Henceforth, as game-players on the stage of life, we keep acting out our specific role-identity and settle comfortably into the whole world of illusion and pretense. This is the basis of our conditional nature. It is an *act* which is difficult to change because it is subconsciously equated with survival. It shows up through many Psycles case histories.

Homeostasis: As explained in Chapter 2, life is fundamentally homeostatic: it responds actively to various environmental forces to maintain its integrity and equilibrium. It reacts to keep an even keel, to "balance the books," to survive, to keep on keeping on.

Because with Psycles we have a more inclusive view of life, we are adding new dimensions to our earlier definition of homeostasis. We say that homeostasis is also conditional and psychosomatic. Life maintains its integrity not only as a body, but as a body-mind. Patterns of identity and belief are also involved in this balancing act.

Thus homeostasis and the maintenance of equilibrium are transferred to the conditional self-image or identity that we described, and a threat to this self-image is now considered as a threat to life. Action and reaction is in accordance with this model and in manners which are designed to maintain this model.

We say that homeostasis is also *confluential*. (A confluence is the flowing together of two rivers to form one river.) There is an innate cooperativeness among people which goes far beyond our ordinary ideas of how living beings interrelate. The vast subconscious communication which is revealed by Psycles case histories will be discussed in Chapter 12.

Underlying the spring-back efforts in homeostasis, may be the physics of *magnetic tension*. I believe it is more than an analogy to

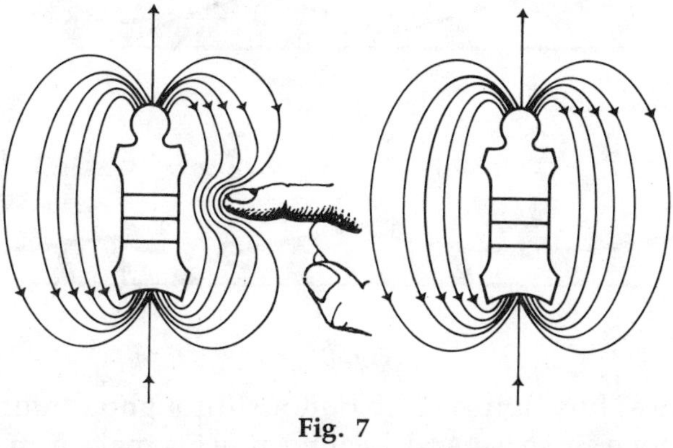

Fig. 7

say that the stimulus-response process derives from the fact that magnetic fields resist being deformed, and spring back to their natural spherical lines instantly when the distorting force is removed (Figure 7). That is why healing is sometimes so incredibly fast.

Reciprocal functions: Throughout living bodies there can be found pairs of nerves, muscles, organs, and organ systems which are in balance but opposed to each other. They are mutually antagonistic or *reciprocal* in their action. As with eye muscle pairs, for example, the stimulation or activation of one means the automatic inhibition or deactivation of the other. Give a plus sign to one, and the other is a minus, in three dimensions. As with seesaws, when one end goes up, the other end goes down automatically.

Various examples of such pairs of structures and functions are listed on page 310. In the same appendix is also a chart of the reciprocal functions of the autonomic nervous system and their target organs.

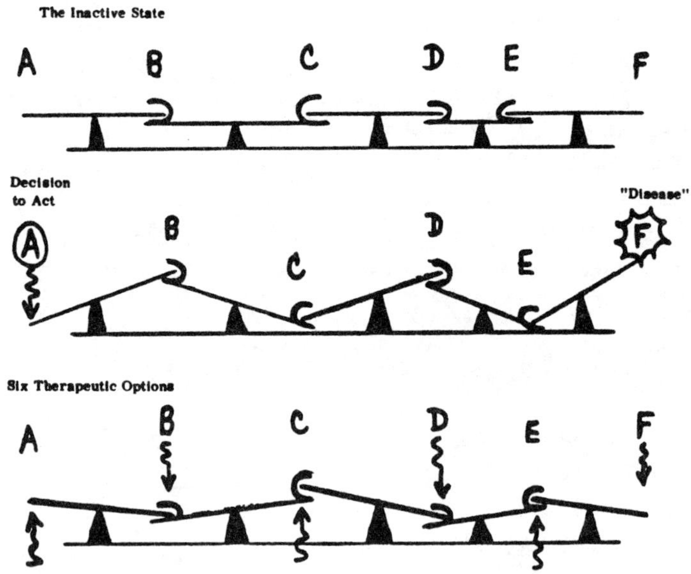

Fig. 8

Reciprocal functions are ubiquitous throughout life. Without this knowledge, disease and healing must remain a mystery.

Why does it exist? Where did it come from?

One of the most revolutionary clues, overlooked for decades, lies in the embryology of the spinal cord. There, in one picture, is a secret of secrets, the bridge between cell divisions, mirror-image symmetries, *and* all the reciprocal functions—laid out for all to see. We won't describe it in detail here (See Appendix G). All we want you to know is that our revolutionary new picture of the physics of life (not its chemistry), is firmly grounded in biological facts. And when we suddenly recognize the magnetics that permeates our being—we find our boundaries reaching out into the cosmos!

The Genesis of Disease

Depending on the self-image, and who we think we are that needs self-preservation, and when this image is threatened in whatever manner, there is the decision to respond in that manner or fashion which is most appropriate, logical, and necessary from past experience to restore it.

There are thousands of ways in which we modify tissues and distort structures and functions. The responses are whole-body reactions, involving the entire 3-D symmetrical network of cells down to the level of the final target cells. The distortions, or "diseased" responses, are done with precision. They are specific, meaningful, and motivated. The ways in which we respond, from cell to cell down the chains, is by activation and inhibition which is reciprocal from cell to cell, involving alternating plus and minus signs (states of activation).

This includes all the spectrum of nonoptimum conditions known as "diseases," tensions, attitudes, feeling-states, "hang-ups," phobias, accidents, bizarre behaviors, "weaknesses," and all manners of death.

Aside from death itself, these distortions become "locked in" in due time, to become automatic and seemingly autonomous. But they are still governed by the conditional self-image—which can be changed!

This is the essence of our conditional nature and the foundation for the genesis of disease. In the maintenance of self-images lie the roots of human aberration and suffering. This model of how we work and the genesis of disease is illustrated in the left-hand picture on page 54.

Now that we've touched on body structures, symmetries, reciprocal functions in 3-D celluar-chain networks, and homeostatic identifications, let's return to our classic case of Judy's throat hemorrhage and its healing.

What Really Happened to Judy?

We shall now retell the story of Judy in more detail, because it can serve to illustrate fundamental principles as we discuss the general mystery of disease and healing.

In Colorado Springs, on a Sunday evening late in 1952, and with friends and neighbors sitting around a kitchen table, a vivacious 18-year-old brunette named Judy complained of a throat hemorrhage which had bothered her all day.

"It's funny," she said. "Can't figure any reason for it, but I've been spitting blood all day, though otherwise feeling fine. It just happened out of nowhere."

"When did it start?" I asked.

"Oh, it's just been all day."

"Did you wake up with it this morning?"

"No. Must have started about 10:30 A.M."

"Going back 37 hours—did you have any troubles Friday evening about 9?"

Her breath caught in Judy's throat, and tears came into her eyes. "Oh God, I don't know what got into Mama!" And she stammered out a story of total rejection. "She said I was no damn good, and nothing good would ever come out of me! And no matter what I said, it just seemed to get worse and worse. She went berserk!"

We listened quietly for some time, just letting it flow. Then—

I said, "Judy, let's suppose that the two events were related, like cause and effect. And that you did the hemorrhaging because of how rejected you felt Friday evening, and that it has meaning for you—something you learned long ago, that might make things right. Can you remember any time in the past when you first had a throat hemorrhage, and what might have happened?"

Through the dim haze of the past she remembered. "When I was about three, I almost died of a strep throat. I was spitting blood by the pint. Mama said that the doctor tried everything—that was before the days of antibiotics—and there wasn't anything more he could do. He said that the rest was up to me. And you know, that night, hour after hour, Mama was at my bedside, praying for me to live. It was almost as if she had never loved me before that night. I remember it. . . I lived."

There was the equation, deep in her heart! It was the memory of an association between that throat feeling and being wanted and loved by her mother. And when, fifteen years later, mother exploded with hate and rejection, and she searched deep within herself for a solution, there was that equation. By making her throat bleed, on purpose, she recreated a feeling-state which was associated with mother-love. Subconsciously, she resolved the rejection, at a time-lag of 37 hours.

Remember this story. Because within all of us there are similar memories of good and bad associations, which we use for psychosomatic homeostasis—to solve problems, by balancing the books within our subconscious. This case is just one of many. It illustrates a fundamental principle which will become clearer as we continue.

But there was more to it than that!

By turning on the throat hemorrhage, Judy was also triggering a reminder to her mother of what happened many years ago. Deep in the mother's heart was the memory of how she had treated the child as an infant, how the child almost died, and how she had felt terribly guilty, and how she had prayed for life. There's nothing lost to memory. We may think we have forgotten things, but there is much proof that it's all still there in the unconscious, fully active.

So Judy was not only recapturing the mother-love feeling, she was also doing a sneaky control mechanism, triggering the mother's guilt. This is actually a very common way in which people control each other.

As we discussed earlier, to diagnose her ailment or "disease" medically, she could have had dozens of tests, blood-vascular, gastrointestinal, or even neurological, in an effort to find something "wrong" with her. And among the hundreds of chemicals, enzymes, hormones, or secretions in her blood, urine, or tissues, some of them would be found abnormally high and others abnormally low (as part of her changed state). As a result there could be found many possibilities for "causes" to be interpreted as biochemical disturbances or even hereditary weaknesses. Furthermore, scrapings from her throat might easily produce some little protein structures which reproduce as "viruses" for a diagnosis of viral infection as the "cause."

But we know the real cause. Judy was *doing* the bleeding on purpose, as an act of will (subconsciously) because of its meaning to her. She made a decision to rectify mother-rejection by purposively recreating a physiological feeling-state which occurred 15 years earlier. We would insist that there was nothing "wrong" with her. She didn't *have* a "disease." She was just *doing* what she wanted to do, for reasons which were appropriate for her. And what she did was manifest through a network chain of reactions, 37 hours later.

In Judy's case, between the homeostatic decision to act and the final total response of throat hemorrhaging, she had to bring about, however subconsciously, a whole chain of reactions in reciprocal fashion in a three-dimensional network of body structure. These included hormone changes, neuromuscular signals and contractions, cellular activations, and dilations of certain blood vessels—each with their associated electrochemical modifications which go with body chemistry.

Her whole body was involved!

Judy was no freak. We are *all* doing similar things to ourselves, for similar reasons, and therefore the same principles hold true for all of us.

The Mechanisms of Healing

With knowledge of the psychogenics of disease, the 3-D network structure, and the reciprocal functions between cells, healing can now be easily understood as the switching of states of activation.

Between the original decision to respond and the final tissue changes, there are many places (cell junctions) throughout the network of cells. At most of these junctions one can be "poked" with various forces or energies to "change the signs" in order to alter the response, reverse it, or turn it off.

And, as in a chain of seesaws, with ups and downs linked together in series, there are many *ways* in which an "up" can be pushed down, and a "down" pushed up, to change the whole pattern of activation.

By stimulating, shocking, counteracting, blocking, poking, interrupting, and thereby affecting *any link* in the causal *chain*, the whole state of activation can be switched around. Effective forces include various chemicals; drugs; electrical, thermal, magnetic, electromagnetic, light, or mechanical pressures; hormones; foods; substrates; radiations; surgical penetrations or excisions; proteins; enzyme catalysts; ions; minerals; salts. . . . Practically anything goes.

That is why there are so many different therapies and techniques of healing which *work* in spite of all the different theories. At one point or another in the 3-D network, they act to "change the signs," or reverse the state of activation, and suddenly the patient is "well" (i.e., he or she quits the whole response pattern).

In Figure 9, the clenched fist symbolizes *any* disease or trouble which a person is *doing* on purpose. To *heal* it, it is necessary to induce the fist to unclench.

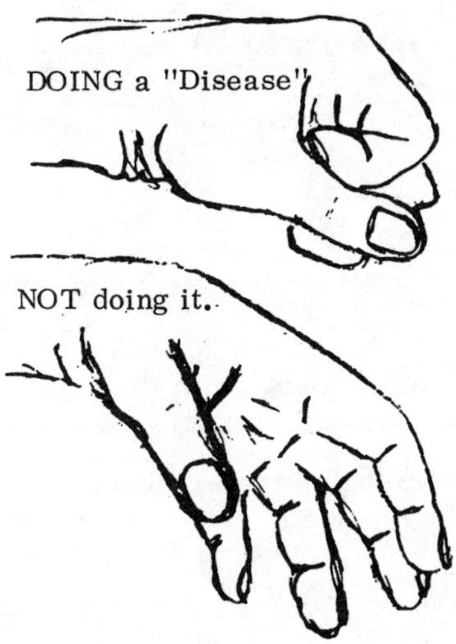

A change of *intent* will do it,
or a karate chop to the neck
or a tickle in the armpit
or a needle in the elbow
or an electric shock to the wrist
or a burning cigarette in the palm.

Fig. 9

"*All* of these "therapies" could induce the hand to open. Therefore *any* of them could turn off the "disease" and be a "cure."

Now, do you see why there are so many therapies with so many beliefs about disease and healing?

In Figure 10 we illustrate both the genesis of disease and the

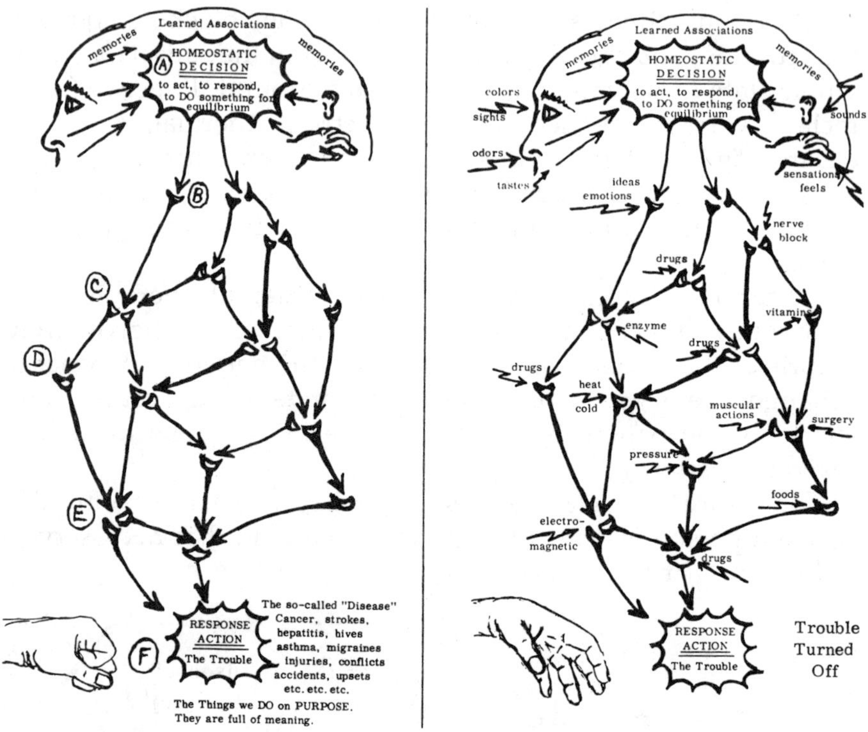

Fig. 10

Disease

We deliberately modify our body structures and functions—

On purpose—
In accordance with the needs for the maintenance of self-image, set, conditional associations, etc.

Through a three-dimensional network of neuro-anatomical structure, down to every cell of the body.

and

Healing

Therapy consists of—
a) whatever may induce us to change our minds/motives—and quit doing it, or
b) the application of countermeasures/efforts/forces (of many kinds) at one or more cell-junctions (throughout the intervening network) to turn off the aberrant response (a mechanism involving reciprocal innervation).

healing process. There are numerous options for changing the states of activation. At (A) itself (where there was the decision to respond in the particular manner to maintain the self-image), there can be either a change of mind (image) through insight or a shift of orientation or a change of motives and goal. Or, the decision can be opposed directly with a countersuggestion with or without hypnosis, such as by invoking divine authority.

At the various links in the chain of responses (B, C, D, E) the state of activation can be reversed by the various physical stimulations listed above. And at (F) an opposing force may be applied directly to the final response, such as for instance, in a state of fever, dunking the entire body in ice water.

Most systems of therapy involve more than one change-function, so that physical and psychological methods may augment each other.

Ideally, there should be a change of self-image with the associated intents and motivations. Otherwise, without addressing the primary psychogenic cause, we may "relapse." That merely means that if our deep reasons for doing the distortion haven't changed, we may decide to do it again.

And that is what *Psycles* is all about.

Disease and Healing in a Nutshell

1) What we *are* is a conditional self-image imposed on a 3-D network of cells in which the pattern of activation is reciprocal between cells in chains.
2) What we *do* (when this image is threatened) is to respond by distorting our structures and functions in ways meant to maintain that image. Hence, most of our troubles and diseases are responses designed to preserve that image.

   If this be the case, then, for healing, health, and happiness—
3) What we *can do* is
   a) USE whatever techniques you find effective for deactivating the troubled distortions. (There are thousands of techniques for reciprocal sign-changing).
   b) But more importantly, *act* to *de-condition* that self-image which underlies all your troubles. That's your salvation! because—

4) What we can *be* is Un-Conditional. For *there* lie the answers to the mysteries of metaphysics. In the unconditional are manifest all the ultimates of human potential.

With this final note, we are now ready to enter into the marvelous discoveries which Psycles has made into the genesis of many internal distortions which we call diseases.

4 Revolutionary Solutions: From Asthma and Strokes to Muscle Spasms

In the last chapter we discussed a new concept of "disease" and "healing" in which most internal troubles are seen as behavioral problems to be solved, rather than illnesses to be cured. From the standpoint of our 2,000 Psycles case histories, our view would be that there is usually nothing technically "wrong" with us. Instead, we are doing our troubles on purpose, for logical reasons, and with incredible precision down to the modification of individual cells within our bodies. We don't *have* diseases; we *do* them.

Now in this chapter we will start with the first of a series of troubles involving internal distortions, to show not only *why* they are done, but also *how* healing can occur.

Asthma

The Asthmatic Cyclist

A young man came out from Pennsylvania to the dry desert heat of Arizona because of his asthmatic problem. The change seemed beneficial: for several months he had not had any severe symptoms. But one evening at about 10, he returned from a motorcycle ride, wheeled the cycle up a ramp, "accidentally" let it fall, expended great effort in raising it, and immediately went into an "attack."

It was a serious one. The choking, strangling, and near suffocation were relieved by injections of drugs which he carried with him and which friends administered. For him, the reason for the attack was the effort put into raising the fallen motorcycle. But for us, after questioning, the deeper cause was a letter he had received 37 hours earlier, which conveyed a severance of relations with his fiancée back East.

We will discuss the probable *root* cause in a moment.

But consider the activities which he went through in order to give himself the plausible "reason" for the attack. This meant considerable subconscious maneuvering, perfectly timed, so that when the appropriate time came for the explosive homeostatic response (the need to feel suffocation, etc.) he had the "reason" in the situation he had created. He had "bought a ticket" to permit this action.

Before exploring the psychological reasons for the genesis of this condition, let's review briefly the rationale for asthma to be found in medical source-books.

The Medical View of Asthma

First we exclude the asthma like symptoms associated with congestive heart failure (cardiac asthma), pulmonary edema (lungs filling up with fluid), pulmonary embolism, endobronchial TB, bronchogenic carcinoma, and Loeffler's syndrome in diphtheria.

Then, coming to bronchial asthma, we have two categories: a) extrinsic, atopic, or allergic asthma, presumably due to a hypersensitivity to airborne materials, and b) intrinsic or infective asthma, presumably due to bacterial infection of respiratory areas. Actually the two categories are diagnosed almost interchangeably, with the latter thought to be allergic although the true antigen hasn't been found, and both categories supposedly influenced in frequency and severity by such "secondary factors" as weather, fumes, fatigue, endocrine changes, and emotional stress. As stated in one text (*Merck Manual*,1972,p. 543): "Emotional factors and stressful situations can play a large role as a secondary factor. It is important to uncover disturbances in the home or at work. Psychotherapy is seldom successful. A trial of one of the tranquilizers may help." Various drugs alleviate the symptoms.

What does the asthmatic person actually *do* to his lungs? The lungs are distended, with thick tenacious mucus plugging the smaller bronchibles; the mucosal and basement membranes are thickened, the alveoli distended, and septa often ruptured; mucus glands show hyperactivity; air is obstructed by the narrowed lumens and excess secretion; there is wheezing and prolonged expiration.

While most doctors have very little psychological orientation, and therefore treat asthma as allergic, infective, genetic, etc., with some vague "emotional stress" considered as a "secondary factor," there are some psychosomatic doctors with views coming much closer to the real causes we find with Psycles.

Dr. Rosenbaum, in *The Mind Factor* (Prentice Hall, 1973), views most respiratory ailments as "disguised crying," with asthmatic behavior as a repressed cry, for a rejecting mother's attention and love. For him, asthma is a function of separation anxiety, with an unresolved dependency on mother. We will show how close he came to it with our next case.

The Communal Asthmatic

Philip was seven years old, living with his parents in a commune in northern Washington. In this case we have a whole series of homeostatic reverberations to one big trigger incident, and the rich symbolism of it reveals clearly the relationship of asthma to the trauma of birth. After expulsion from the womb and the cutting of the umbilical cord, the newborn infant must galvanize new throat and lung reflexes, coupled with the termination of the umbilical circulation, or else he strangles. In the womb, he "breathes" through the umbilical cord; when that is cut, he would suffocate, except for a whole new set of reflexes associated with lung breathing.

Here is Philip's story in chronological outline:

Friday, 8–9 P.M. (4/27/73). Crisis meeting, in which his parents confront the rest of the commune with their decision to sever relationships, sell out, and leave the group. To Philip,

The Communal Asthmatic

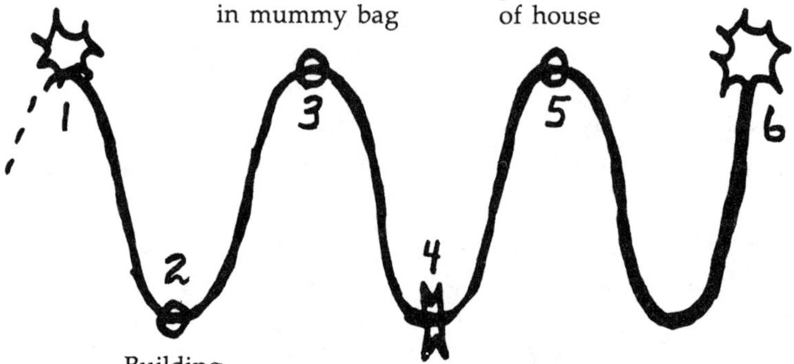

1. Friday, 8–9 P.M. 3 days before
2. Saturday, A.M.
3. Saturday, 8–9 P.M.
4. Sunday, 9 A.M. c. 37 hours before
5. Sunday, 8–9 P.M.
6. Monday, 9–10 P.M. Severe asthmatic attack

apparently, this was like cutting the cord, terminating a womblike security, and a transfer to an unknown world "out there." Because . . .

Saturday, A.M. He was busily erecting a triplet of playhouse apartments (wombs).

Saturday, 8–9 P.M. With special permission, he slept out in the *"fort"* in a *mummy bag*.

Sunday, 9 A.M. He had his initial asthmatic symptoms: hoarse throat, coughing. And, he finds an *empty nest* in the woods, which he brings in, to take to school.

Sunday, 8–9 P.M. He returns home to find himself *locked out of his house* (door was locked because they were laying linoleum) and has a tantrum.

Monday, 9–10 P.M. Severe asthmatic attack, with choking, throat constrictions, tight chest.

Symbolic of the womb are the playhouse apartments, the secure fortress, the mummy bag, the empty nest, and his own home. During those three days in between, he seemed to be reaching far and wide, because of his fear of the impending expulsion from the commune, to find some resolution to the problem, with various symbolic efforts to retreat to the womb and its umbilical security. And all of these homeostatic efforts were on the circadian half-wave. Add them all up, and you have asthma as a function of birth trauma.

How do you talk to a small boy choking?

First of all, don't underestimate his cosmic wisdom. Remember the meeting, and the decision to leave. Discuss how this might feel like being born, and how birth happened. And how the umbilical cord was cut, and tied, and how he had to start breathing through his lungs . . . and how he did, after all, do it and how he survived it! And most of all—love him.

Some day, much later, he will have the tremendous experience of "returning" to the experience of birth, reexperiencing it, feeling the pain of it, through and through, so that it can be deconditioned of its trauma. That is the ultimate and only real *cure* for asthma. And by the time he's ready to tackle it, there will probably be faster and more precise ways of helping people to do it.

In the meantime, for each "attack," just check it out, and make contact with the series of events leading up to it, using Psycles—the circadian half-wave. That can be an effective solution. Strange things happen, when the wheels start going round and round inside . . . when you start to see the reasons why.

To go back to the Arizona case: it would seem that the severing of the relationship with his fiancée was equivalent to being cut off from mother, thus triggering the same trauma of birth.

Interestingly enough, the dean of American allergists, Dr. Murray Peshkin, in 1966 recommended what they called "parentectomy," or the removal of the patient from his parents, for serious and persistent cases of asthma which don't respond to drugs. The parents are, of course, the continuous reminders. They reported that "cures" frequently happened within 72 hours of getting the child into another environment.

Sneezing, Coughing, Vomiting, Diarrhea, and Rejection

We all wish to be loved, accepted, and allowed to be the way we are. When someone is hostile and shows by his angry behavior that we don't deserve to exist, we feel it. And this feeling of rejection, of being invalidated, ousted, or threatened with symbolic annihilation, has a rather common mode of homeostatic rebalance.

What many of us do is a self-ejectory action, as though turning inside-out, with sneezing, coughing, spitting, and, in more severe cases, vomiting or diarrhea. These appear to be homeostatic for what we call "the rejection syndrome." Although some happen at the usual 37-hour lag, more often they seem to happen on the first half-wave of the circadian rhythm—at about 12+ hours.

One Sneeze

I picked up a hitchhiker one day who asked me what I was involved in. I handed him a copy of a newspaper story on my research. After reading it, he erupted with one sneeze. It was about 10 A.M.

"Did somebody reject you, put you down, last night about 9:30?" I asked.

"As a matter of fact, yeah." And he broke into a wide grin.

"That's how it works."

"But what good is it to know that?"

"Well, with this simple tool of analysis, you can be more self-aware, and not pass over too quickly something that was important, like a clue as to where you felt attacked or invalidated. Something about your self-image, and what you believe about yourself, and what you might do about it."

Several Vomits

On a quiet Tuesday evening I was in my study working, when suddenly, at 9:30 there came cries of anguish from the room of my four-year-old son Brian. We found him in a puddle of vomit. After being cleaned and comforted, he went back to sleep peacefully, but at 10:30 P.M. he again woke up vomiting.

This double incident of vomiting might have resulted in anxiety, frantic calls to a doctor, or other emotional displays with adverse effects on Brian. But instead, our first response was to look into the events of our lives. For instance, what happened this Tuesday morning at 10—nothing? Monday morning about 9:30—nothing much? And how about Saturday evening 9–10?

Ha! A pattern emerged quite clearly. Brian was vomiting *for* me, but on a lag of about 72 hours. Or put it this way: I had not faced up to a severe rejection (a double incident) at that time or I might have vomited myself Saturday evening. But, repressing my feelings, I developed hot pains in my stomach which had been persisting off and on for the past three days. And now Brian, being tuned to me subconsciously, erupted to relieve *our* tension.

One great function of Psycles, then, is the alleviation of anxiety. We did not get frantic about his sudden and mysterious vomiting. This was good for him, and good for us. We had every assurance that there was nothing "wrong"—just a healthy expulsion of stomach contents. This was homeostatic for feelings of rejection. The books were balanced, so to speak. And I thanked him, for being my teacher.

In another typical example of the rejection syndrome, a young lady artist in San Francisco had drunk a little too much at a party and become too loud and aggressive. She touched on some sensitive point of one of the male guests who suddenly turned on her in fury, and cut her down with a vicious verbal assault.

The next day while out driving, exactly 12.5 hours after the incident, she suddenly felt nauseated, stopped the car, got out, and vomited. It was not the twists and turns of the road, nor the exhaust fumes, nor anything she had eaten. It was homeostatic compensation for feelings of rejection.

We often manufacture "good reasons" for our troubles, because it is easier and more comfortable to say that it was that damn—whatever—that we ate that didn't agree with us. After all, who wants to remember a painful incident? But the cause is usually *not* what we ate. Note this, however. Clever as we are in concealing painful truths from ourselves, we sometimes engineer "good reasons," for instance, by going out and deliberately eating some garbage—just so that the nausea which we are about to do will be logical! We are master game-players.

One young mother was frantically calling for medical help because her six-month-old baby was vomiting nonstop. Backtracking from the initial vomiting, 37 hours, took her back to a crisis in which her drummer husband was ousted from a rock band. When it was suggested that the baby was just responding to the father's rejection (being part of a family "gestalt"), the connection between the two events brought sudden peace to both the mother and her baby.

Coughing Fits

A case of violent, continuous, hacking coughing—without redness or soreness of throat—followed 37 hours after an incident in a grocery store, in which a neighbor accused the subject of being stingy. In recalling and discussing the event, she remembered how her obstreperous younger brother had frequently accused her of the same trait during her early childhood.

One sudden coughing fit was triggered by a "serious run-in with the boss" about 12.5 hours earlier. Another sharp coughing bout was preceded 37 hours earlier by his wife's caustic comments on his TV viewing habits.

Diarrhea

Millions of cases of diarrhea are probably a simple function of the dating game, as for example in the following incident. A young executive arrived at a party, greeted his girlfriend with an affectionate hug, only to be pushed off coldly without explanation. He spent the rest of the evening feeling turbulent emotions of jealousy, wondering whether she had shifted her affections, and with whom. Thirty-seven hours later, he had one single and total diarrhea, without subsequent symptoms.

We are continually selling ourselves short, by pretending to be dumb and innocent victims of circumstances. But this is not the way it really is, at all! Part of the self-liberation adventure is in knowing this. With Psycles, even the lowly sneeze can lead us one step closer to freedom.

"Colds"

Colds are another complex of symptoms which take us right back to birth and the prenatal environment. In the hundreds of cases which we have checked, we have almost invariably found—back either 37 hours or three days—incidents involving loss of security, expulsion, sudden abandonment, or other reminders of birth. And here again, the symptoms are a reproduction of the prenatal state or birth itself with, in many cases, the blood-vascular constrictions connected with the first breathing reflexes. The "cold" appears to be an effort to return to a condition of the past, in which there was meaning of a survival nature. We *do* colds on purpose.

Sometimes, as a result of the sudden insight from awareness of the trigger incident, the symptoms may vanish just as suddenly. Sometimes they hang on for a long time in spite of our knowing something about the problem through that insight. This would seem to indicate that the desire for that feeling is very strong and remains necessary for lack of alternatives and solutions.

Colds occur routinely with such events as job lay-offs, firings, change of residence, and other sudden reminders of birth, in which there is something equivalent to being thrust from the security of the womb out into a painful, frightening, or cold world. That feeling of security may even be terminated by the loss of a friend, running out of money and being in financial trouble, or being threatened with such a loss.

It is predictable that the symptom complex known as "colds" might be much less common among those infants whose births represent the least trauma: where the mother is mentally and emotionally prepared, relaxed and confident (as in the Lamaze method), and where the lights are dim, it is quiet, and the temperature is warm; and also where the infant is immediately placed in contact with the mother's body, and remains in close and loving relationship.

In the system advanced by the French physician, Dr. Frederick Leboyer, the newborn infant is transferred into warm water, and gently stroked and massaged until an ecstatic smile spreads over its little face. Every effort is made to reduce the shock between the internal and external worlds, and to make the transition as peaceful and comfortable as possible. By contrast, the usual hospital methods can be considered as criminal, with babies tossed around like so much meat, in sterile nurseries, lonely, crying for their mothers.

In a simple classroom experiment, a goldfish was traumatized by letting it flop out of water until nearly exhausted, and then returned to the bowl for further observations. At about 37 hours later, while everyone was expectantly watching the goldfish swimming unperturbed, the experimenter (yours truly!) suddenly began to manifest cold symptoms, starting with a sharp sore throat. Apparently, the out-of-water goldfish had triggered my own birth trauma!

The Squelched Teacher

When she suddenly felt a painful congestion in her throat, a teacher throught it meant the beginning of a cold infection. But as it turned out, 37 hours earlier she was in a board meeting of a private school, prepared to make some strong statements about a serious problem the school faced. She was forbidden the opportunity to speak and sat there fuming in anger.

The throat problem carried several elements of motive. There was rejection, yes, but also vocal repression and rage, all blended into that one response.

Some people actually prefer to think of themselves as innocent victims of circumstances or as helpless pawns in a cosmic chess game. But full self-responsibility is not really that hard to accept. Wouldn't it be better to *know* why things happen? Because in this case, the minute she connected with the board meeting and its frustration, there was no more sore throat!

Hives—and How We Can Bust Out of Our Skins

The skin is a very special organ. It marks the boundary of a unique territory—the me, my body, the I. Even though "we" may extend energetically far beyond such boundaries, there is this territorial limit which encloses our physical/molecular structure. Any unauthorized penetration is resisted with a whole-body, alarm reaction. For this, the skin is the first line of defense.

Now what sort of events and relationships might cause us to erupt with hives? Why might we *do* hives, on purpose?

The Tardy Beauty Queen

Sitting with her family at the dinner table, peacefully having supper at 7:15 on a Monday evening, was a recently chosen beauty queen.

"Ooh, mama, look!" she cried, scratching her wrists. "It's starting to itch, and my face too!" And within minutes there were ugly welts on the back of both hands. Red splotches began appearing on her cheeks, and by 7:30 P.M. her lower lip hung swollen in an irrepressible pout.

No obvious trouble could be seen in the events Sunday morning, 37 hours before. She had remained in bed, asleep, until about 9, had then dressed without incident for a new date for an outing in a sports car. But Monday morning, a scant 12 hours earlier—ah, there was the trouble! She was scheduled to be picked up at 8:00 for her first important modeling job with a prominent advertising firm.

Awakened at 6:20 A.M., she was fixing her own hair. Mama remained in bed, angry, because her daughter had returned late the night before and, without the proper preparations, might again be late this morning. The hair was rebellious. Frustrations piled up, until a final SOS went out to mama at 8:00, to get the final touch. The party was kept waiting for ten minutes, while both mother and daughter went through the motions, with repressed anger. Neither exploded—until that evening at 7:17, when it came bursting out through her skin, as hives.

The Engineer on Vacation

One warm Sunday morning, a research engineer was departing for a vacation trip to Florida with his family. During the last-minute loading of the car, their irascible neighbor, with whom they had been feuding over various problems, including an overhanging tree, leaned over the fence and began prodding old wounds.

But this was not the time to get caught up in another fight, so he buttoned his lips, tightened his jaws, and let the caustic remarks and snide jabbings go unanswered. All he wanted was to get out of there as quickly as possible.

Late Sunday evening, on the drive south, he pulled to the side of the road to take a short nap. This was 34 hours after the departure from home. About two hours later, he awakened, covered with hives. He had erupted from head to toe. Would this have happened if, 36 hours before, he had cut loose from the guts and let his anger out at his neighbor? We don't know.

The Itchy Hindu

One hundred and forty miles north of the city of Bangalore, in south India, there was a Hindu cloth merchant who was recently expelled from African Kenya. He was visiting the ashram of the famous Sai Baba, worshipped by 50 million Hindus as the avatar of the age and renowned for his omniscience and miracles. The merchant wanted personal advice where to set up business again and what to do, but the days passed, as he waited in vain to be beckoned into private audience. Feeling increasingly impatient, he decided to write out his request, sit in the front line during "darshan" time (the public appearances), and verbalize his request for help and advice as he handed him the note.

As the critical moment approached, Sai Baba turned in a different direction, as if deliberately avoiding him, to pass on through the crowds without the contact the merchant wished. Thirty-seven hours later, the Hindu gentleman was itching, suddenly, from head to foot.

"If Sai Baba was like your father, to you, could you be angry at him?"

"Absolutely not! I could be angry with my mother a thousand times, and she would forgive me a thousand times. But my father? Never! That was forbidden. Anger toward him would be severely punished."

The sequel to this incident was equally interesting. A few days later, he procured the services of a Brahmin priest to perform a ceremony for his recently departed mother. But he concluded that the priest did an inferior job and charged too much, and as a result he proceeded to vent his anger against the priest! We conclude that this outburst of anger against the priest was a direct outcome of the original frustration at darshan time with Sai Baba. First the repressed anger, then the eruption of hives, and finally the overt expression of the anger itself at a different but acceptable target, all happened with Psycles timing.

Eye and Ear Problems—and How We Sometimes Resist Seeing and Hearing

From our studies it would seem that essentially every cell of our bodies is under our control, and every function can be distorted or deformed—on purpose. This includes our five senses as well, and they may or may not be painful.

Conjunctivitis from Bad Scenes

During the early years of this research, while I was associated with Dr. Gordon Beckstead at the Psychological Research Foundation, Inc., in Phoenix, Arizona, there was a period of time when I spent every other night (all night) on a small mountaintop in the middle of the desert. It was an exciting and joyous effort toward self-liberation, in which I found my isolated peak before sunset, meditated on the sun setting, spent the entire night in a variety of self-searching disciplines, meditated on the sun rising, and returned to the office.

One early morning, just before sunrise, I witnessed a car streak along a road about a mile away, then leave the road and roll over about three times in a cloud of dust. Fifteen minutes later, I was at the scene of the accident. The highway patrol was already on hand, and there was a body lying on the road, covered with a blanket, with one hand exposed. On that dead hand were several flies. At the time of this observation, there was a slight feeling of revulsion (I had never witnessed an "accidental" death before) but no conscious awareness of any problem within myself with respect to that event.

But 37 hours later, early evening of the following day, I suddenly felt my eyes itching, began to rub them, and within minutes had a full-blown case of "conjunctivitis." Going back to that scene in my mind's eye, and looking at that body, ended the irritation, and the apparent effort to not-see.

Earaches from Bad Phone Calls

Arriving one evening at a friend's home in Laguna Beach, California, I found my host with an agonizing pain in his right ear. At about 5 that afternoon, he dove to the bottom of a swimming pool and then surfaced with what felt like a burst ear drum, with intense pain—only in the right ear.

With his wife's help, we reconstructed the events backward, each 12+ hour interval, and found three very bad phone calls which were in violation of his whole power role as a probation officer. They occurred at circadian half-wave intervals both at the 37-hour timing, and, double that, at three days plus two to three hours. The pain subsided rapidly, and by the next morning his right ear felt entirely normal again.

Chances are that the right ear drum was not burst, but it felt like it. Looking back at the phone calls, it was as if he were making great efforts, within the muscles of his ear, to counteract or cancel out the sounds of those calls. It was as if his right ear was resisting vigorously those contradictory messages.

What were the three phone calls?

One was from an important person who wanted a favor he couldn't grant. One was a critical call from his wife in which there was a bad connection and some vexation. The third was from a female client in a divorce case who wanted to bypass him to talk to his wife. All three were more complex and frustrating and annoying than can be shown in a few words.

Psycles later came the counterefforts within his right ear. But notice how he made them "logical" by diving to the bottom of a swimming pool? So that the excruciating pain had a plausible "cause" in the immediate situation?

And from the Wrong Call

In San Francisco there was an earache resulting not so much from a bad phone call as a wrong one. An artist awoke at 2:40 A.M.

with a very painful earache in her right ear. Tracing back a day and a half, it was shortly after 1 P.M. when she returned to her apartment, hoping for a phone call from her boyfriend with whom she had had an argumentative rift.

The phone rang. She rushed to pick it up, with hope and excitement. But it was from someone else—a call totally unwanted. She was furious and disappointed. Thirty-seven hours later came what seemed to be an inner effort to push those unwanted sounds out of her ear, coupled with the whole tension and anger.

Heart Problems—and How Meaningful They Are

I watched a miserable program on TV last night—a special on heart attacks. Summarizing the latest medical findings, it focused almost exclusively on the seven "risk factors" to which one must pay attention in order to avoid the higher probability of being a heart-attack victim.

And I asked myself: how can all those doctors seem to be so ignorant? Why must they act so blind? Surely, they are at least somewhat aware of the fact that bodies have heads and that heart attacks certainly must have something to do with people's personal problems? The word "psychosomatic" was not mentioned once in the entire 90 minutes! It was as if heart troubles just happen to you in statistical manner, like roulette, totally un-personal.

Then, assuming that they really do know, somewhere deep within, the question became: why do they pretend to be so ignorant? And the answer came: it's because people don't want to know; most people prefer to be out of touch with reality on this subject; they're afraid that it might be too painful to recognize the reasons underlying the troubles. Thus most people continue to pretend to be innocent bystanders. They choose to be unaware, and the doctors go along with it.

Now let's change all this! It's really not so terrifying.

I know, and many others know, with confirmation from our many case histories, that people do their heart attacks for reasons. Whether they die or not depends on whether they intended to die or not. There are many other things people do with their hearts beside turning them off suddenly when they wish to die. We can cause them to speed up or slow down, cause pains, or just jump erratically. Actually, we cause strange heart behavior in the same way we wink, spit, or twiddle our fingers. It's that simple (or mysterious!).

The proof lies in the fact that the decision/trigger can usually be found in an earlier event, at the predictable Psycles time-lags.

For example: an ambulance rushed out to a home in Phoenix, Arizona, to pick up a patient who had had an asthmatic attack. With oxygen he was comfortable and talkative. But the strange thing was how that man's wife cried when they took him away. There are many asthmatics in this area, and most "attacks" are survived.

But she knew, apparently, deep in her heart (subconscious) that he had made his decision. And sure enough, 37 hours later, he stopped his heart in the hospital and died. That was one of my early case histories. I was working for the Grimshaw Mortuary for a short while, almost thirty years ago. I drove the ambulance, picked him up, saw his wife cry, talked with him in the ambulance, and delivered him to the hospital. Two days later I picked up his body and saw the time of his death on the death certificate.

Some people flaunt all seven risk factors outrageously and still live to a ripe old age. In other cases, heart troubles seem to go in families. This has given rise to the notion that there must be a genetic factor. But that again ignores all these case histories, showing identifications among other motives.

The Fluttering Heart

A female patient came into the clinic complaining not only of a fluttering heart, but an inability to swallow—as though she had a lump in her throat—and difficulty in breathing. A quick check back 37 hours by the receiving nurse brought into focus a sharp quarrel she had with her husband in which she held in her anger. After reviewing the incident, she ended up chuckling, realizing her inability to "swallow the situation," and her symptoms vanished.

The Irregular Heart

Into the same Phillips Clinic in St. Louis came another woman patient with an irregular heartbeat, feelings of faintness, and other symptoms of acute anxiety.

The receiving nurse-therapist made special note of the timing of the initial symptoms, and then asked what had happened at a time 37 hours earlier. The patient exploded with a story of near-catastrophe involving her infant son and her husband's costly tools.

"I was accidentally locked out of the house, and there he was, my two-year-old Jimmy, strewing all these tools around on the back porch. It took me a half-hour of wheedling and pleading to get him to let me back in. If anything had happened to those tools—my husband would have killed me!"

She cried in telling about it, promptly became clear of symptoms, and left the clinic healed.

And Another One—on Me

I suspect that most readers would privately confess that they have experienced irregular heart beating on more than one occasion, but kept it secret bcause of such questions as insurance premiums and the basic fear that something might be "wrong."

One of my occasions followed 37 hours after viewing a TV special on Evel Knievel, the daredevil cyclist. I saw him doing incredible death-defying feats and, in slow motion, some of his injurious falls. Now, everyone watching the same show would not necessarily manifest a jumpy heart 37 hours later. This happened to be a "thing" with me, regarding death. So for me it was significant, and triggered a deep response. Knowing the primary triggering cause, however, was sufficient for me to be assured that there was nothing "wrong" with my heart. I was merely doing the heart dance, on purpose, until I recognized what was disturbing me.

The Soviet Commissar's Heart

In case you wondered, the same Psycles time-lags apply to the Russians. Employed in 1977 as a biologist by the University of

Washington, I spent two months in the Bering Sea alone with 92 Russians on a 285-foot Soviet fishing vessel. It was my job to observe their fishing and compile a detailed report which would help the U.S. Government in its management of this enormous fishing resource.

One of my routines, besides the scientific work of monitoring the catch, was to spend two hours every evening as guest in the Commissar's cabin. One evening he confided to me that he had serious heart trouble: his heart had been beating quite irregularly for several days and he could hardly sleep. There was no doctor on board, as the young female physician, fresh from medical school and on her first voyage, had been so terribly seasick that they had to ship her back to home port in Siberia.

As diplomatically as possible, I probed the several days preceding his symptoms and discovered why he was so terrified. He told me of two sailors who had brewed some rotgut down below, got roaring drunk, and said some very bad things to him (such as probably threatening his life!). He felt obliged to ship them back to Siberia on the transport which periodically took their frozen catch. This talking *plus* some digitalis which I had in my old World War II medical kit switched his heart back to normal, and he was truly grateful for the relief.

The Bursting Heart

What is more central to life than the heart? Our musical, poetic, and literary language is full of references to the heart in connection with all kinds of interpersonal relations. Our hearts are "bursting with love" or bleeding, or breaking, or hardening, cheating or opening up; our hearts are fearful or brave; we are soft-hearted or warm-hearted or cold-hearted.

All of these conditions can be manifested physically, with the right conditional response.

A patient who had recently undergone open heart surgery managed to recollect some of the events which preceded his initial "heart attack," scanning it in accordance with Psycles timing. The reasons for the heart attack and the subsequent surgical repair became quite clear.

A 55-year-old father of five sons fell hopelessly in love with a married woman on a Sunday morning. Monday evening (37 hours later) he called her up, his heart "bursting with love," pleading with her to throw caution to the winds and enter an affair with him. Though she said she'd think it over, the signal was probably negative. Wednesday morning (37 hours later), no information. But Thursday evening at 8:30 (another 37 hours) he called her, only to be crushed with an overt rejection. Halfway up the stairs from the basement phone, he had his heart attack, which consisted of a large ballooning of the left ventricle. (His heart was literally bursting.)

Several months later he had open heart surgery which showed a three-inch balloon, which was cut away and the ventricle sewed up. He had colored photographs of the two large three-inch pieces of his heart which were excized. And thus his broken heart was patched up and repaired.

With insight alone, might the ballooned ventricle spontaneously return to normal structure? Considering the incredible control we apparently have over individual cells and tissues, this is a possibility. But there are limits to homeostasis. And in this case I would rely on outside surgical intervention to repair the damages.

Just remember that people make decisions to stop their hearts, for reasons. And should you be in the presence of such an event, figure backward in time. Ask where they were and what they were doing at such and such times. Spell it out. Try 37 hours and then 75 hours (three days and three hours), give or take an hour or so. Expect to hit paydirt, i.e., some critical event. Don't let some foolish person try to shush you up. In those moments you may crack into that decision, open it up, and reverse the whole course of that life! You might save someone you love.

More of methods, later.

Hepatitis—and How We Even Modify Our Livers on Purpose

The liver is a large and exotic organ-system with many marvelous functions which no one really understands very well. For one thing, it is definitely involved with fundamental questions of survival. It is concerned with life or death decisions.

The Hepatic Surfer

My son Brian was a dedicated surfer. So religiously devoted was he to the sport that even through the dead of winter (it's not always sunshiny in southern California) he would be out on the ocean waves at the crack of dawn, proceeding to school much later.

A very cautious person, he had passed up contest after contest, feeling that the time was not quite right, and he wasn't ready. He was deeply involved in surfboards, shaping his own forms, sometimes doing his own fiberglassing. He had visions of continuing in the industry as a professional. So when he finally entered the first contest, it was an important decision, and an important event on which his whole future depended.

He lost. The waves were not right. Many small circumstances "conspired" against him, and he failed. It was a catastrophic defeat, deep within himself.

The first symptoms came 37 hours later. They worsened. After three days a medical diagnosis of hepatitis directed him to spending the next six months flat on his back. It was a bleak outlook. I call this the "belly-up syndrome" because a defeated dog, turning his belly up, is guaranteed life. His opponent wll not touch him from that instant.

It was into this scene that I returned home one day to realize—good grief! my son was in serious trouble!

But when we laid it all out time-wise, pinned it on the contest, opened up his expectations and failure, spelled out future options—it cracked wide open! Within 40 hours he was off the whole hepatitis syndrome and back on the ocean, with the beautiful body bouncing back to its old energy again.

Hepatitis among the Devotees

Hepatitis is a common occurrence among the foreigners who go to India to spend some time with famous gurus. They go there with high expectations of divine intervention to transform their lives miraculously. There are great experiences, great "highs," and then there are great disappointments when their noses are rubbed in some of the conditional roots of their identities. Sometimes, when they find deeply immense barriers between where they are and where they would like to be, it can seem catastrophically impossible to surmount. Time after time, 37 hours after such confrontations, the first hepatitis. The belly-up syndrome. For the most part they survive and continue the psychological adventure with renewed enthusiasm.

Migraines—and Other Headaches Don't Just Happen without Good Cause

In church one morning I was talking to Goldie when she mentioned that she had awakened early with a strong headache, right across the eyebrows.

"So what happened Friday afternoon, late?"

"No problems at all."

"How about a flash on anything in the last three days or so which might be especially binding, frustrating, like a 'mexican stand-off' tension?"

She thought for a moment and then said, "How about Saturday morning. That's where it is!" Saturday morning for the first time she met her exhusband's wife—the woman who "broke up my marriage." She had always wondered what she was like, and now she was forced into seeing her as well as her ex carrying their baby. She noticed, very queasily, that she was almost a spitting image of her, tall, blond, slim. Like looking into a mirror. The whole complex of emotion in her head became the headache across her eyebrows, 24 hours later.

I get lots of my case histories in church.

Another headache belonged to Lisa. Hers was on awakening Saturday morning at 7. First, I asked what happened 37 hours before, Thursday evening about 5:30. She got a phone call from her mother announcing that mother and father were coming to spend three months with her, through the summer, and then go off to Arizona for the winter. She lives with her three children in her parents' former house and doesn't get along with her mother very well. The mother is arthritic, tense and critical.

Dennis was attending a college class in geology one Thursday afternoon, when at 3:15 he suddenly had a powerful headache combined with an overall "upset" feeling, so severe that he had to leave the class. Wednesday at 3 A.M., 36 hours before, he was with a girlfriend in a motel when the police came banging on his door looking for someone else.

Virginia's strong headache came 37 hours after a heated argument with her grown son regarding her treatment of the other children in the family. Kathy's headache, combined with aches and pains in the neck and shoulders, began exactly 37 hours after a phone call in which she was ousted from a business project she was jointly planning with two other women. Most troublesome was the fact that woman A called to say that woman B (who was Kathy's close friend) couldn't go through with the association, and Kathy wondered why B couldn't have told her face to face.

As you can see from these few cases, the possibilities for headaches are almost limitless. All kinds of interpersonal problems involving conflict of interest, feelings and understanding can underlie tensions within the head. Just exactly what happens is unknown, inasmuch as no one yet knows exactly the mechanism of "pain." It's one thing to say that it involves blood-vascular constrictions, but the whole subjective aspects are quite mysterious.

Migraines have similar root problems, but just why it should feel like an icepick through the brain, or why it may involve the electrical brainstorms ("scintillating scotomas") is again mysterious. Of course one can point to the semantic concept of "I can't see (why or how, etc.) . . ." as what might trigger visual aberrations. In other words, we may be acting out the inability to "see" (understand) something.

First, I'll summarize three migraines which happened after the typical 37-hour time-lags, and then I'll review at length one which had more unusual timing.

A doctor reported to me that one of his patients, a Miss S.S., had severe migraine-type headache, combined with nausea, vomiting, and pains in the left shoulder and knee. When he questioned her about events 37 hours prior to the symptoms, this woman was feeling great resentment at having to entertain a friend's daughter on a Sunday afternoon, yet pretending otherwise.

Nita's migraine began while she was instructing her pupils in ballet about 10 A.M. on a Saturday. It was Thursday at about 9 P.M.

when her husband left her stuck at home with two children, while he went out to a choice seminar which she very much wanted to attend. She was furious.

Charlie was a handsome 28-year-old who still fancied very young girls. His migraine turned on 37 hours after being socked in the eye by a 14-year-old teenager who didn't appreciate his attentions.

The following case seems to have a distinct four and a half day lag between the sharp conflict and the visual brainstorm, or scotoma. There were, however, several in-between events, at 37-hour multiples, which seem to be significant in the sequence.

A young man attending university was still living with his mother, and one of their problems was a running argument whether he should keep his room door shut and how this might contribute to the stale smell which his mother objected to. The son kept insisting that when the furnace blower turned on, the largest portion of the volume of air in his room was circulated down through the furnace, and anyway, his room *didn't* smell, and it was an unreasonable and fetish-type hang-up of the mother. The argument went on and on.

Returning from a vacation trip of several weeks, he found his room had been thoroughly deodorized in his absence, including a most horrendous crime: his favorite Mexican blanket, which gave brilliant color to his room, had been put through the wash and was sadly muted, its colors now drab.

"You've ruined my most precious possession!" he shouted. "I wish you'd get off my back about room smell." At that moment he felt a deep surge of violence as he never had before. There were the usual claims and counterclaims which peaked within a brief quarter-hour or so, and that was the total incident. But during the next hours, up to about midnight on this Thursday evening, he felt greatly agitated, with a silent dialogue continuing in his brain.

Saturday noon, about 37 hours later, he drained the oil from his car and refilled it with a heavier oil, including two cans of STP, to try to offset an impending engine breakdown. A rod was about to go. This may be symbolically related to the tension he felt, and meant an effort to reduce the pressure.

Sunday afternoon, another 37 hours later, he found himself involved in deep philosophical discussions on fundamental secrets of life. This involved strenuous intellectualizing with many problem-solving efforts within the brain, and could be considered as relevant to the series of events, like an effort to break through an impasse, a sort of "mexican stand-off" situation.

Finally, four and a half days later, or three multiples of 37 hours, he suddenly began his visual aberration, the "brainstorm." Starting with the realization of a visual blind spot, he closed his eyes and saw jagged lightning streaks near the pupil. Gradually, during the next 15 to 20 minutes, this jagged streak expanded out to the periphery of vision and vanished. Within half an hour it was gone.

How do we interpret this?

The specific visual image, the brainstorm, remains a mystery to me. But, in my extensive conversations with this young man, I began to feel that the deepest significance lay in some very early experience in which his feces were rejected by his mother, with revulsion, and that as a newborn infant he tuned into a feeling of rejection which was equivalent to annihilation.

In subsequent years, there may have been a series of reinforcements in which body odor became an issue, as a system of control, which was quite subconscious but effective between mother and child. In other words, the mother could gain control of the child by threatening to eliminate his odor, and the child felt it as a personal threat, and responded with agreement to *be* controlled.

Of course this is highly conjectural, and there could be a wide variety of analyses and interpretations. Suffice to say that events themselves and the four-and-a-half-day lag in response is what interests us here, however they may be second-guessed. We shall have more to say about more extended time-lags later.

Neuritis—and How We Can Generate Neurological Troubles on Purpose Too!

From scores and scores of miscellaneous case histories of aches and pains, shooting pains, sizzling pains, spasms, etc., I shall select just a few to illustrate typical trigger incidents. Making contact with such underlying reasons can end a great deal of anxiety which millions of persons are feeling concerning the possibilities of something being "wrong" and the expenses of treatment and cures.

The Toe Ache

Susie was an expediter in an aerospace factory. One day, just before noon, walking along and minding her own business, she suddenly experienced a sharp and persistent toe ache in her right big toe. The whole toe was shouting in pain. Unable to walk, she found a bench and sat a while reflecting on the whys and wherefores. As it turned out, 37 hours earlier, she was in bed with her husband, engaged in one of their recurrent arguments about women's rights, and she was resisting an overwhelming impulse to kick him out of bed!

It bears repeating that sometimes the connecting, in memory, with the trigger incident can be highly therapeutic. In this case the pain went away within minutes of the insight.

The Sizzling Foot

One of my writer friends took a temporary job with the city of Seattle as a typist while struggling to get his magnum opus published. One of his assignments brought him to the building department one morning where he was scheduled to begin days and days of transcribing the recording of a public hearing.

First there were several hours of frustration, because the responsible person was absent and there was no suitable Dictaphone with foot pedal available. After he finally got the equipment set up, he found that the transcribing was going to be a horrendous task. His foot went back and forth on the pedal trying to get word for word what various persons were saying into the various microphones. Speech was often slurred; there was an excess of distracting noises, and there came one lengthy and hard-to-understand witness who said something unintelligible (but seemingly very important) followed by a loud cough right into the microphone.

He tried desperately to get what was said, backing up over and over, and each time there was the "blah . . . blah . . . blah . . . COUGH." After about the tenth loud cough blasting into his ear drum, this fellow got up and said, "That's it!" He threw off the earphone, disconnected himself, and just left. "That's not for me!" He quit with, at this time, a sharply rising headache.

Thirty-seven hours later, he was sound asleep.

Another 37 hours, he began to notice some aches and pains in his neck, which went away shortly.

Another 37 hours—four and a half days later, came the first slight and brief neuritis in his right foot. Lasted just a few moments.

But after one more day—exactly five and a half days after the Dictaphone crisis—sitting in a restaurant at 10:30 P.M., there was a sudden and sharp sizzling pain in his right instep that was so intense and shocking that he jumped and slammed his right knee onto the underside of the table. Four hours later came another sharp sizzle, lasting less than half a minute. Now came the analysis, and the pinpointing of the trigger incident. But that didn't end the problem. Additional shocks came on days 6, 7, 13, 19, 20, 31, and 33. And that was the end, without any further recurrence of the sizzle.

The Muscle Spasms

Cindy's trouble was sudden, sharp, and painful recurring muscle spasms in the trapezius muscles, "like labor pains," with shooting pains running up her neck. What brought them about seemed quite clear to her. It was about a day and a half earlier that she had the first full realization that her daughter, age 15, had left home permanently. She had had a great many problems with her daughter. She was very mature, at 15, and uncontrollable.

But fancy having "labor pains" at this time!

Some Leg and Ankle Pains

A clerk in a wholesale office was walking along when he had a sudden pain in his calf. He continued walking, and found that instead of increasing the pain, it felt better. In a stooping position, the pain increased, and then remained constant with additional walking. At about 37 hours earlier, he had a feeling of intense disgust with the situation at work, his employer's indecisiveness, and a strong urge to be away from the whole scene.

The Would-be Olympian

Saturday morning, between 9 and 10, a normally joyous and spontaneous woman of 52 was discussing her deep indecision regarding a trip she had to make. "I used to be so much more spontaneous, with a sense of rightness," she lamented.

Saturday, 9 P.M., she was out to dinner with a machismo Greek who was dramatizing that type of spontaneity and simplicity. Nothing complicated about him! He was straightforward, naive, proud, acting without deliberation or conflict.

Sunday morning she said, "If I had my life to live over again, I'd start out being a champion athlete. Here I am fifty-two, and not famous for anything!"

Sunday, 9–10 P.M., she was watching Olympic Games on TV, featuring downhill skiing and speed skating.

Monday, 9:30 A.M., she awakened with sharp pains in both calves, in their lower portions, and running down the main tendons. She wanted to see an acupuncturist to find out what internal organs might be amiss, such as her kidney, spleen, or liver.

But with a little Psycles scanning, her anxieties about internal organs were quite relieved. The pains in her calves were the efforts to *get moving* and quit her indecision. The interpretation was not imposed on her; as soon as we touched on the series of events, she "saw" it herself. That was her own conclusion.

Now note the timing involved. There were "significant" events at each of a series of five half-Psycles, from Saturday morning to Monday morning. Though 37 hours is a very frequent time-interval for interesting cases, it should be remembered that a full analysis would take into consideration each multiple of the half-wave (12.5 hours) of the circadian rhythm.

5 Hypoglycemia, Diabetes, and the Whole Blood Sugar Problem

From various reports in the late 1970s, there were estimates ranging up to 70 million people in the United States suffering with problems attributed to high or low blood sugar, with 300,000 deaths due to diabetes and 50 million hypoglycemics complaining of dozens of different symptoms.

This represents an enormous multibillion-dollar drain on the American economy and, according to a report of the National Commission on Diabetes, there was "widespread resignation about the inadequacy of the present therapy."

If You Are Hypoglycemic

There is almost no human condition that cannot be reversed, almost no state of disease and debilitation that cannot be overcome. The human body is incredibly resilient. It is marvelous. It is backed by millions of years of ability to survive!

Naturally, if you have lost an arm or a leg or a whole organ, there is little likelihood of your growing another. So there are certain limitations to this marvelous ability intrinsic to life.

But hypoglycemia is not like the loss of an arm or leg. It is a deformation, a distortion, which is reversible. It can be changed. It is not a permanent acquisition. It is readily modified. You can quit it! This is our promise.

You remember those photos of the victims of Nazi brutality who barely survived the gas chambers of Belsen and Dachau and other extermination camps. They were bare skin and bone, many unable to walk. Eyes sunken in hollow sockets, with gaunt looks, every rib standing out sharply against a thin and stretched skin. Hands like claws, looking more like Halloween skeletons than human beings.

And yet, within a few months, many of these same creatures were fleshed out in radiant health, showing no evidence of their near brush with death. How can such amazing transformations occur? What manner of structure is it that can manifest such radical change of dimensions and appearance? What secret image lies hidden within that grotesque horror, such that a transcendent beauty can emerge in total contradiction to itself?

You, too, no matter how much you have suffered, no matter what painful distortions you have done to yourself, no matter how long you have stooped and grovelled, moaned and groaned—you, too can be transformed into a thing of beauty. And you deserve it! You deserve to be healthy and happy, creatively fulfilled. Have you condemned yourself to be otherwise? Way down deep, don't you know you can change it?

Those who have fallen back on the assumption that hypoglycemia is hereditary have done so, we believe, only because they have been so frustrated, *thus far,* in their efforts to change it. Unable to see any possible avenue of escape, they felt defeated. They gave up hope, and surrendered to that hopeless idea that the condition is genetic, and cannot be reversed.

But it can. And the reason it can is that we have discovered its fundamental mechanism. And with this new knowledge, it can be conquered! Courage!

The Hypoglycemic Engineer

My first confrontation with hypoglycemia was in 1965 while working in a laboratory of the aerospace giant Autonetics in Anaheim, California. There, a case erupted right in front of me and gave away its secret.

It was a Monday morning, May 10, at about 11:45, when a chemical engineer suddenly collapsed, without warning, at his desk nearby. With head down, arms folded, shuddering and shaking, he fumbled in his pockets, pulled out several sugar cubes, maneuvered them clumsily into his mouth, and within five—ten minutes quickly recovered so that he was quite "normal" again within the quarter hour.

"Well!" I said, "That was something. Hey! What happened?"

"Hypoglycemia," he said.

"I've heard of it, vaguely, but tell me about it."

"Well, hypoglycemia is super-low blood sugar, that goes with hyperinsulin, or too much insulin. Zaps all the sugar out of your blood, and suddenly you feel drained of energy, and in the extreme, like a state of shock."

"Amazing!"

"It all happens every now and then, unpredictably. The medical treatment is a special diet, with high protein and fat, but very low on carbohydrates. No sweets or goodies."

"Would you mind trying a little experiment? For example, can you remember what you were doing Saturday evening at about 10:45?"

"What's so special about that?"

"Oh, I'd like to check out a new theory, from a study of biological rhythms. Did you have any troubles that evening?"

"Troubles? Hell, no. Just watching TV. No troubles . . . everything fine."

"Can you remember what you were watching?"

"Whatever was on . . . you know."

"This might be very important. See if you can remember the program."

"OK. Ten forty-five . . . yeah . . . now I remember. It was a movie about a plantation in South America. Charlton Heston played the lead. He got bad news of a long black line of army ants heading his direction, destroying everything in their path. So he set up three lines of defense, with ditches, trenches with water, and hedges to set afire. The ants breached the various perimeters of defense, and closed in on the final inner defense line. There was fire, flooding, and total catastrophe, wiping out seventeen years of effort."

"How fantastically paranoid!" I exclaimed.

Gary looked at me through defensive and suspicious eyes. "Are you implying that that movie had anything to do with this attack?"

"Could be."

Before suggesting how it might be that this experience could trigger a hypoglycemic attack 37 hours later, and to get a broader perspective on the whole problem, we will outline the medical picture, to see what bodily changes are involved.

The variety of symptoms which have been associated with low blood sugar reads like the Who's Who of Suffering. But lest you be influenced by this list, remember that some people have low blood sugar and *none* of these.

The Medical Model

A. The syptoms (the way we feel and act) may be:

1) Sweating, flushed, chilly, numb, pale, weak, faint, hurt all over, tired, drowsy, fatigued;
2) Dizzy, trembling, disoriented, restless, nervous, apprehensive, hyperactive, hypersensitive, "freaked out";
3) Increased blood pressure, pulse rate changes, heart palpitations, emotional, uncoordinated, depressed, hungry, moody, negative, easy crying;
4) Colitis, hemorrhoids, receding gums, hair falling out, cracked tongue, headaches, double vision, memory loss, think-speech confusion, alcoholism, bed-wetting, schizophrenia, blind staggers, spouse-beating, and even killing;
5) Seizures, convulsions, coma, death.

Some doctors add a long list of other diseases which may also be "caused" by low blood sugar, such as hay fever, asthma, rheumatic fever, peptic ulcers, epilepsy, multiple sclerosis, addictions, and neuroses.

According to the medical model, all these various symptoms can be *caused* by low blood sugar. We think it better to say that these symptoms have (sometimes) been associated with a state of low blood sugar, but that the condition is actually a response to deeper *real* causes, which we will show presently. But to continue the medical model:

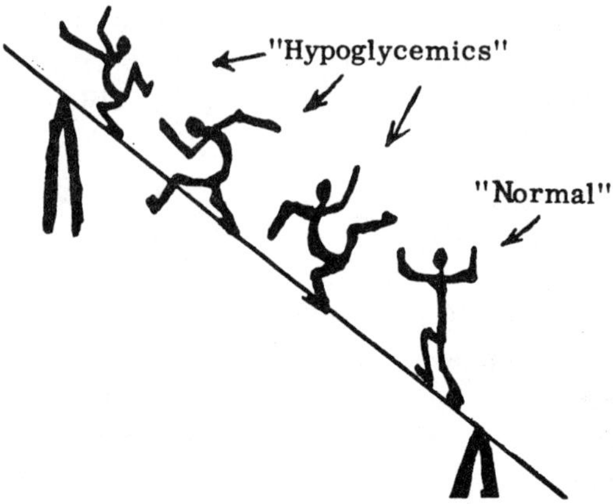

B. The low blood sugar ("causing" all the above symptoms) may, in turn, be because of the following possible conditions:

1) Temporary states, such as severe muscular exertion, or poor nutrition during pregnancy and lactation;
2) The islet cells of the pancreas overproducing insulin because:
   a) spontaneously overreacting to glucose absorption;
   b) an islet-cell tumor, or just a proliferation of these cells;
3) Liver troubles which alter the glucose production and glycogen balance;
4) Adrenal cortex troubles from infection, tumor, starchy degeneration or hemorrhage, impairing hormonal control over the pancreas;
5) Pituitary troubles from infection, tumors, atrophy or necrosis, impairing hormonal control over the pancreas.

If now we ask what causes *these* upsets, imbalances, atrophies or overgrowths in these other organs, we arrive at where the buck stops. Because most medical thinking concerns headless bodies and is not very holistic, they arrive at two alternatives: a virus, or you're born that way.

Treatment, therefore, is limited to:

1) Surgical removal of the hypersecreting tissues, or otherwise treating the infected glands (very few of these).
2) Corticosteroid hormones to stimulate the residual functions of the system (a paliative).

3) Diet of high protein and fats, with low carbohydrates and sugars, and carrying sugar cubes for emergencies. (Suitably triggered, the hypoglycemia still happens).
4) Psychotherapy for the emotional disturbances (without much hope of really changing the basic condition).

It is generally agreed that the medical model for hypoglycemia is quite discouraging. Excluding the very few "organic" cases which can be helped by attention to the other glands (liver, adrenals, pituitary), this means that the 30 to 50 million hypoglycemics can only be "managed" (and poorly at that) by a stringent diet.

What Controls Blood Sugar?

What we have is an interconnecting system, with checks and balances, which maintains the status quo of blood sugar level in the body. This system includes the islet cells of the pancreas, the adrenal cortex and medulla, the liver, the anterior lobe of the pituitary gland, and the thyroid gland.

They are like a complex set of seesaws. With sugar and starch intake, and a rise of blood sugar, the the islet cells of the pancreas secrete a protein hormone, insulin, which promotes the conversion of glucose by the liver into glycogen. The excess sugar is thus stored this way in the liver, muscle, and nerve cells. If there is a great excess over body needs, such as may happen after a meal rich in sugar and starches, the kidney excretes it over a certain level (the renal threshold, which is 160–170 mg/ml of blood).

Normal range of blood sugar is 80–120 mg/ml blood. If this level drops below 40 mg, adrenalin is released from the adrenal medulla, promoting conversion of glycogen back to glucose for tissue use. If the glycogen reserves are exhausted, proteins and fats are broken down and converted into glucose sugar, and the excess nitrogen excreted.

During fasting, blood sugar levels are kept up by another islet cell hormone, glucagon, which signals the reconversion of glycogen to glucose. Thus glucagon and adrenalin both push up glucose levels, in opposition to the function of insulin.

Pituitary gland hormones are also involved. With one hormone it controls the adrenal cortex, and with another, it has a direct effect on the rate of glucose utilization in the tissues.

The adrenal cortex produces still other hormones which are injected directly into the blood; they aggravate diabetes, hyperglycemia, promote an increase in liver glycogen and blood sugar, while leaving muscle glycogen unchanged; they also stimulate breakdown of proteins.

And finally, the thyroid gland produces a hormone which stimulates the rate of metabolism, increasing basal metabolism by as much as 50 percent; it promotes tissue breakdown of proteins, accelerates the use of fats, depletes liver glycogen, and enhances the rate of glucose absorption.

Thus you see something of the complex, seesaw (reciprocal) systems involved in blood sugar levels, and why it is no simple matter to control it. The whole body is involved. And that includes the head, thought- feelings, emotions, and the whole body-mind system.

The Focus of Medical Research

We can't begin to summarize the enormous research efforts which have gone into the problems of blood sugar, whether diabetes or hypoglycemia. Most of it is irrelevant to the revolutionary breakthrough we are presenting here. Because they generally treat bodies on a purely chemical basis, medical researchers have focused at the level of organs or tissues. It is at this micro-level of the *effects*, rather than dealing with real *causes*, that we find the great expenditure of money.

Among the areas of attention are the following:

1) Among the endocrine cells of the pancreas, there are beta cells producing insulin and alpha cells producing glucagon. Efforts are made to modify or control these cells and their productions.
2) Surgical implants of new islet cells themselves.
3) Use of another hormone, somatostatin, to control the insulin and glucagon productions.
4) Searches for viruses which might cause or trigger hypoglycemic conditions.
5) Research into the whole subject of immune system; some would even consider that diabetes (the insulin-dependent variety) is a disease of autoaggression.

6) How to control or modify the ways in which hormones act or interact with specific molecular receptor structures of membranes (and hence how drugs get into cells).

These are just a few of the areas of research—all of which avoid the whole psychological basis of hypoglycemia, and the revolutionary answer that lies in Psycles.

We can't enter into a lengthy discussion of the complex relationship between hypoglycemia and diabetes (high blood sugar). They are certainly close cousins to each other.

Although all diabetics have high blood sugar, some have no symptoms where others experience frequent urination, thirst, or hunger. Some produce little or no insulin, whereas others produce a great deal, but release it at the wrong times or fail to respond to it. Nearly all seem to be subject to the complications of diabetes: heart disease, blindness, cataracts, blood vessel damage, nervous disorders, and kidney damage.

Just as we saw the many reciprocal functions involved in hypoglycemia, so it is with diabetes, and a newer, more holistic analysis taking into consideration the fears and insecurities of the patient will undoubtedly bring it all into perspective, with similar solutions as we will show.

Lacking though it may be in the deeper psychological insights of the problem, there is the now classic book *Body, Mind and Sugar* by E.M. Abrahamson, M.D., and A.W. Pezet, published originally by Holt in 1951, with subsequent reprintings, which details many theories and analyses of the problem and its treatment. He recommends the Harris diet for hypoglycemia, and many people have found this high protein and low carbohydrate diet helpful—at times.

The Psycles Breakthrough

Now let's return to the Autonetics engineer who saw a paranoid movie late Saturday night and, 37 hours later, manifested acute hypoglycemia. What new interpretation can we place on it which might help *solve* the problem?

When we scan the symptoms associated with hypoglycemia, with Psycles in mind, we see a person who is deeply afraid. And further, if we look at the physiological events which happen we see one of the most ancient survival mechanisms being activated. When confronted with a survival threat, animals must either escape or fight for life, and there comes into play a very important series of responses. From the adrenal glands comes a jet of adrenalin. This causes the liver to immediately release blood sugar needed for violent activity. The increased sugar triggers the pancreas islet cells to produce a jet of insulin, without which the sugar cannot be utilized for energy in the cells.

Sudden fear, therefore, galvanizes the body for violent action. Throughout history and evolution survival threats meant action. But today we are suspended in a state of inactivity, neither running or fighting, and we're in trouble.

Millions of people saw the same movie. Why didn't they all react with hypoglycemia? The reason is that we are each unique in our past experiences. The painful things which happened to me are quite different from the troubles which happened to you. We're all "balancing the books" within our subconscious minds, but because we each have a different set of experiences, we respond differently in what we think will do the job—survive.

For Gary, something in that terrifying scene of swarming black ants carving a path of destruction reminded him of something similar in his deep past—when he was very small and helpless. Whether fire or flood or ants or even "aunts," he felt threatened. His whole body was demanding violent action for survival.

But what could he do? Put his fist through the TV tube? Jump up and down and scream at the ants? He was probably quite unaware of his deep agitation. Excited, yes, following the drama. But disturbed? Not as far as he could put his finger on it. But his survival response, blocked at the time, erupted 37 hours later in an insulin jet which depleted his blood sugar. If at the instant of the TV crisis he had jumped up and engaged in violent muscular activity, he might not have had the troubled response. But that is pure conjecture.

The "Cure" Is Not Always Welcome

During these years of exploration, we have had contact with close to a dozen cases defined as hypoglycemic by the six-hour glucose tolerance test. Strangely enough, not all hypoglycemics want to take the steps necessary to solve their problem.

Many become convinced by the literature that their condition is intrinsic to their physiology, that they are "just that way"—which may be defective, biochemically, and that it can't be helped, and they just have to put up with it. Rather than look inside, they probably feel more comfortable with the dietary approach, relying on the power and prestige of medical authority.

The Professor Who Quit It

It was a great day for Psycles when the first severe case of hypoglycemia, medically diagnosed with the six-hour glucose tolerance test, was restored to normal living.

As a professor in a western university, he had a wife and two children, and suffered the dismal routine of eating a separate diet of hamburgers and salads, while the rest of the family enjoyed gourmet cooking. Even this dietary regime did not prevent him from frequent bouts of acute symptoms, when he almost passed out with near-total depletions of energy. He was resigned to a tortured life, never knowing for sure whether he would be feeling genuinely well at any particular time.

What makes a person decide to try something different? What degree of desperation brings about a turning point in a person's life? What distinguished him from others who reject change and the effort involved? He began to keep a diary to help him remember events, and began to check out the acute symptomatic occasions. The first time he tried it was a bull's-eye! Almost to the minute, 37 hours before, there was a critical incident with all the earmarks of a logical reason for what he was doing to himself.

He was insecure. And there was a series of crises having to do with his job and his relationships with others on the faculty. There were confrontations with deans, staff meetings, some pointed hostilities of several other professors who were critical of some of his more radical ways of handling classes. All played a role in triggering acute crises in his hypoglycemia.

By using the 37-hour calculations, he actually found out how and where he was threatened and frightened. He felt the paranoia at its genesis. After a half-dozen incidents, each relating to an upheaval in job security, he came to the point where he felt the problem was conquered!

He had to be honest with himself to do this. It took courage. He came to realize, experientially, the conditions underlying his own feelings of insecurity. Each time a crisis occurred, he became more aware of what was happening and its possible consequences. He began to pay more attention to what he was doing internally to generate the fearful feelings. These feelings were not "just happening" to him as though he were an innocent victim, but states that he was doing, on purpose. And more and more he knew that he could do something about it.

Within four to six months he was effectively over the problem. For one thing, it meant a return to the normal dinner table, enjoying his wife's excellent cooking. His life was renewed.

It isn't easy for most people to consider themselves responsible for their problem. The suggestion that they may be bringing it upon themselves because of emotional or psychological problems involving relationships with other people, their fears, or insecurities may be too difficult to accept. Because of misunderstandings about the word "psychosomatic," they probably wish to avoid the idea that they might be just a little "crazy" if it were "all in their heads."

But once we realize that we're *all* psychosomatic, and that this is universal, and normal, then we need not have that misconception.

The Hypoglycemia Divorcée

Betsy was in the middle of divorcing her husband of many years. She still had two teenagers at home, wanted to go into real estate and be independent, but was still feeling rather insecure about her solo venture.

Sunday she left the two younger boys with her husband for the day, while she adventured with a friend. At 6 P.M. she phoned to find out how they were. Her husband replied that they had been a problem, and in response to his complaints, she accused him of being like a little kid himself. He exploded over the phone. She felt some terror over his potential violence. He had physically abused her on more than one occasion during their marriage.

The next day she called, inviting him to dinner, to smooth things over. So Monday evening was sociable and friendly.

But Tuesday morning, about 7, she awakened with symptoms of hypoglycemia: deeply depressed, no energy, and with dizzy shakes. The acute symptoms followed 37 hours after the threat of violence, in spite of the friendly relations in between.

At first she thought that if hypoglycemia was psychosomatic, the cause was her general feeling of dependence on him. But with the 37-hour backtracking, it pinpointed a specific fear of physical violence.

She had been on a restrictive diet, but when she saw these psychological roots of the problem, she began, cautiously at first, to eat some of the goodies that she had been avoiding and found, to her delight, that there were no physiological repercussions.

Thus the beginnings of another ex-hypoglycemic.

A Word of Caution

We are not recommending that you quit whatever you are doing, medical or otherwise, which may be providing you with apparent relief. We are recommending that *while* you are going about your other ways and means of making life more bearable, you may *also* use Psycles to open up the deeper reasons, to help you find out why you are creating the problem in the first place, so that you may quit it permanently!

One of the classic problems for the therapist is the patient who says, please, solve this problem of mine without touching the rest of me! It can't be done. We can't wipe out one key aspect of the person (hypoglycemia) without touching on other aspects of that personality-structure. Actually, the *you* personality is a complex of many facets and themes interlocking in one self-image, which has been reinforced by many years of trial and error. And it is inevitable that changing one aspect will serve to loosen up and modify other aspects, which all interrelate. This may seem frightening.

Because of this interlocking of the whole personality, even though the problem is effectively solved, there may remain the residues of habit patterns, and new insecurities may revive old responses. But rest assured that any such changes are all for the best in the overall adventure of self-discovery and the gaining of freedom from the painful bindings of the past. It all happens little by little, step by step, quantitatively. Have faith that the solving of the one problem will advance you toward greater health and freedom.

6 The Mystery of Epileptic Seizures—Solved

DOWN through the ages it has been terrifying and unnerving to see an apparently "normal" human being suddenly and unpredictably going into violent convulsions, doing things uncontrollably, and minutes later being normal again.

For the 2 to 4 million epileptics in the United States, it is small comfort to be told that they are in good company: Caesar, Napoleon, Alexander, Lord Byron, Van Gogh, Dostoevski, Da Vinci, Mohammed, Mendelssohn, Handel, Socrates, and so on. Because, after all, *they* couldn't be cured, either. And for the rest, is it not frightening to think that this could happen to just anyone, like you or me?

Fortunately, that era is past. With Psycles, that is all changed; the mystery of epilepsy has been solved; its precision cure is at hand. And although change may be slow, because of the enormous investment in old models, it is only a matter of time before the epileptic tiger is tamed.

It's an exciting moment in history.

What Is Epilepsy?

Most briefly, epilepsy means recurrent "seizures."

Medically, the seizures are defined as "paroxysmal disorders of cerebral origin, sudden in onset and of brief duration, characterized by recurrent attacks involving changes in the state of consciousness, motor activity, or sensory phenomena."

The seizures (or things that people *do*) come in three broad categories:

1) fleeting interruptions of consciousness, compulsive day-dreaming, a short blackout, staring into space, with maybe a twitching limb, lasting 5–30 seconds, with resumption of normal activity as if nothing had happened. These are known as the "petit mal" seizures.

2) falling stiffly to the ground and going into convulsions with body jerkings, possible biting of tongue or incontinence, lasting up to 2–5 minutes. These are known as the "grand mal" seizures.

3) any sudden alteration of consciousness of time, identity, colors, vision, sounds, emotions, disorientation, dizziness, etc., which may be followed by automatic motions such as walking aimlessly, smacking lips, picking at clothes, which may last from 2 to 5 minutes. These are known as "psychomotor" seizures.

A fourth category was added in 1978 by Dr. Walter Alvarez in his book *Nerves in Collision:* the "non-convulsive epilepsy," characterized mainly by sudden violent tempers, rages, arguments or "nerves."

From the Psycles standpoint, of course, all these categories are quite irrelevant, and little attention need be paid to the detailed descriptions of what particular things people do.

Equally useless are the efforts to explain the causes of epilepsy.

The Old "Causes" of Epilepsy

When authorities begin to declare that there may be 200 different kinds of epilepsy with hundreds of different causes, one is justified in being a little suspicious that this is simply a mask to cover up basic ignorance.

It is understandable that the doctor wishes to have his patients think that he knows what he's doing, because faith in the doctor has always been a fundamental premise in medicine. The honest doctor knows that he "treats" the patient as best he can, and the patient cures himself, often without the doctor ever knowing the true cause.

And so with epilepsy.

Officially, "epilepsy cannot be cured." It is considered an unsolved mystery. Technically, the cause(s) are unknown.

There are, however, a host of equally official "causes," the most popular being brain damage (though this is rarely shown) and the random firing of scarred neural tissue. Or, lacking a good accidental referent, epilepsy may be considered an inherited metabolic or genetic defect. This view, of course, provides little hope for a true cure. It may otherwise be thought of as stemming from poor prenatal care, poor nutrition in childhood or later, childhood fevers, infectious diseases or poisons, or more remotely, brain tumors.

In the category of brain damage, it is assumed that there may have been brain injury before, during, or after birth, head wounds such as from a car accident, or hereditary defects in the development of the brain.

These are all suspect, if you recall the illustration in Chapter 3 on "ropes and ladders" thinking, and the failure to take into account the person in the window who wants out!

Treating Epilepsy with Drugs

In line with the incurable brain damage models of epilepsy is the dictum that it can only be "managed." Efforts are made through the use of tranquilizing drugs or relaxation training to moderate the condition and possibly reduce the intensity and frequency of seizures, without abolishing the syndrome entirely.

The epileptic is viewed as defective and handicapped. He is therefore generally barred from driving cars, excluded from most employment, and reduced to a social and economic pariah.

There is a long list of drugs which are used to "manage" epilepsy. Each year brings out new ones. Although Dilantin is the most widely used, there are none which guarantee freedom from seizures. Patients try first one and then another, go for a while without seizures, and then try another drug when seizures occur again, for an endless round of seeking for that magic potion which will bring an end to their problem.

There will be periodic tests to measure the blood level of the drug being used. But despite the maintenance of the prescribed blood levels, seizures still occur, and the futile search goes on. The drugging is often so severe that it drastically impairs the behavior of the patient, bringing on the zombie effect.

The report of the International Commission on Anti-Epileptic Drugs appeared in *Epilepsia* (Vol. 7, No. 3, 1976) stating: "that the picture appears not to have altered throughout this century despite the introduction of the drugs which are so widely used today."

On Brain Surgery

Part of the reason for the popularity of brain damage models may be the despair and frustration coming out of decades of efforts and research which have been essentially fruitless. If something is so intractable—psychotherapeutic efforts have failed as well—then it *must* be something "physical" and "wrong" with the brain, even the possibility of microdamage to just some brain cells.

One very peculiar fact exists: the seizures are random. This has no explanation in the entire life sciences, and seems to contradict the belief that the seizures are the result of damaged brain cells. If there is damage, why can there be normal functioning for days, weeks, months, without seizures? Repair of damaged cells is an automatic function throughout life, and occurs continuously. Besides, there is no evidence that *scar* tissues, as such, should be the culprits. They don't *do* anything besides filling space.

And, most puzzling to brain damage theories is the fact that epilepsy sometimes just quits, permanently, without anyone's knowing why.

Once in a great while, a brain tumor may be found and removed, for beneficial results. But from our Psycles evidence, we conclude that the tumor was a *result* of years of internal tension/conflicts—rather than the *cause.* Just as exercise produces bigger muscles, so neural efforts may generate tumors. Cut them out, and the 3-D network of causal chains involved in the action may be removed, to terminate the habitual response pattern.

New brain scanners can sometimes identify foci of activity in the brain cortex, either through electrical monitoring, or by detection of concentrations of glucose metabolism. Surgical removal of these brain areas can likewise result in beneficial effects—65–75 percent "success" in the termination of seizures.

For some, this may be evidence in favor of brain damage theories. For us, it is not.

The First Epileptic Case History

The first fifteen cases of circadian time-lags in stress-reactions, which was published at the Psychological Research Foundation in Phoenix, Arizona, with my associate Dr. Gordon Beckstead, contained no examples of epilepsy. But the report was read by a nurse-therapist at the Phillips Medical Clinic in St. Louis, and it was she who sent us the first epileptic case. She was hesitant at first to even consider checking out such a bizarre theory, but she thought, "It can't do any harm to ask."

A middle-aged male patient had a seizure on Friday morning just before noon. After calculating 37 hours back to about 10:30–11:00 P.M. Wednesday, she asked him what happened at that time.

He replied that he was watching TV. She asked him to recall, if he could, the specific program. And he remembered. It was a murder mystery.

"Was there anything especially disturbing about it?"

"Well, there was an auto crash in it that shook me up a bit, because it reminded me of an auto wreck that I had shortly before my first seizure."

Aha! Suggestive, at least. Was he reacting to a traumatic reminder of the past, some crisis in his life? She was unable to go much further with this case, as it was not her prerogative to undertake extensive inquiries of this particular patient, and he left shortly after.

But from this first case, she proceeded to check other cases in the same manner, and ended up sending the foundation a dozen case histories which were published. They weren't all epileptic seizures, but included a variety of acute problems.

The Seattle Schoolboy

A bright young boy of eleven had a series of seizures between 9:00 and 10:00 on Tuesday, Wednesday, and Thursday evenings, but not on Monday evening. What happened 37 hours earlier to each seizure? That would be 9:00–10:00 A.M. on Monday, Tuesday, and Wednesday?

That was when he was being dropped off at school. What was so terrifying about that?

There were no seizures for several weeks. He was now taking the bus to school. But suddenly, again, he had a seizure on a Thursday evening about 10:00. This time the stress-cause was pinpointed quite specifically. He was terrorized into the back corner of the bus by a bully.

Then came summer, and no school, and no seizures for almost two months. But another seizure followed an occasion of being bullied by his older brother in the back of a station wagon while returning from a family camping trip.

The pattern was becoming clear: he was a "victim" of bullies, and attention could now be focused on a specific personality problem. Not everyone gets bullied. It takes cooperation, inducement, motivation, a signal from deeper identifications which had a pay-off in the past. Who was his model? His secret identity? And why was being bullied such a survival threat?

But then a strange thing happened.

Just when it appeared that a straightforward counseling program on his bully syndrome could lead to an end of his seizures (with some assertiveness training and karate classes to build up his self-confidence), the boy's mother backed out of the whole program. Apparently, she was personally locked into the problem and derived some important pay-offs which she wasn't quite ready to give up. She was the model. Without speculating on what may have been her pay-offs, there is one general principle which I sense out of many case histories, and it is discussed in Chapter 15 under the section called "Young Martyrs." It is as though the child were doing *for* the mother what she would otherwise have to manifest herself, out of her conditional tensions.

This case is important in illustrating the fact that knowing what to do is not all there is to it. Other persons (parents or spouses) may be deeply involved and need to be considered as a part of the problem to be solved.

The Incestuous Brother

In another case a young male had seizures from about the age of twelve until he left home at about seventeen to live in a faraway state. While away, he had no seizures for about four years. But a visit from his two sisters brought it all on again.

Putting a number of clues together, it seemed as though he was mortally afraid of an incestuous relationship. Whether he actually had sexual relations is not clear, but some combination of circumstances formed this association, in which the threat of death or annihilation went with incest. Away from his family there was no problem. But with the visit, he suddenly reacted again.

The pity of it is that, during these four trouble-free years, he was on the standard medical treatment, maintaining certain blood levels of tranquilizing drugs, which severely impaired his mental faculties. He undoubtedly believed that the constant drugging was what kept him seizure-free!

A Seizure in Mexico

During the early research, one important question was whether the 37-hour timing was the same the world around, or whether it varied with the distance from the equator, as the Foucault pendulum does. (Psycles does not.)

A woman volunteered to keep a detailed diary while living in Guadalajara, Mexico. She kept it faithfully for over a month and much to our mutual surprise, included a "blackout" seizure.

Her seizure was triggered by a grisly scene of a horse, lying at the lakeside near her house, slowly dying. Now, most people seeing a dying horse will not respond with an epileptic seizure. But for her it had some horrendous associations. Many years earlier, her favorite horse, with whom she was deeply identified, snapped its leg in a snake hole, was then shot by her husband, and skinned. The skin of the horse was furthermore hung near the house as it was being cured with alum to be used as leather. She was afraid of her husband and it seemed that she took this personally as an act of violence against her.

High-Frequency Seizures

If a person is having frequent seizures, day after day, it means that the trigger-reminder is in the immediate living situation. It would be like living with a man-eating tiger.

The trigger may be the mother's voice saying certain typical words or phrases which may have been screamed during birth trauma or at some other critical time. Whatever the signal, it now exists as a frequent reminder of a near-brush with death—in the subconscious.

Of course the mother or father or spouse is not consciously triggering the seizures. It's not their "fault" and they shouldn't feel guilty. It would merely be that their voice or gesture is a signal-reminder of a terrible crisis in the past which needs to be resolved.

The immediate thing to do is separate. If it's a child, see if arrangements can be made for him to live with a relative, guardian, or substitute parent for a while, preferably in a different city. Be totally apart, without verbal communication or contact for at least a month, and see what happens. We would expect a drastic reduction, if not a complete elimination, of seizures during that time. If there are a few seizures, now is the time to pinpoint their causes in the new setting, by the methods we have described.

Later, reintroduce the child back into the parents' house, with the instruction to sense immediately any tightening-up, any tensions in relations. Keep a detailed diary, recording hour by hour all the events leading up to several seizures. Now, for each seizure, check back in time to pinpoint the trigger-event, looking for commands, criticisms, naggings, labelings, etc. No one can predict what it might be. Only the circadian timing will tell, and put the case on-track for the necessary problem-solving.

Thus, one unfortunate truth is that parents may not be innocent bystanders in the problem but locked into it by their own deep needs. *Their* role must therefore be taken into consideration and dealt with before the epilepsy can be solved. Otherwise, a permanent separation (without blame) might have to be considered, for the sake of the child.

Even Daydreaming

From the widely syndicated column by Dr. Coleman comes a letter from one of his readers: "I can be in a group listening to stimulating conversation, or reading a book, or doing some form of office work, and I temporarily tune out everything for a matter of minutes. . . ."

The reasons, we believe, can be most certainly found at 37 hours or three days and three hours earlier, bringing insight and peace of mind. Likewise for La Tourette's syndrome, which involves uncontrolled outbursts of profanity by otherwise socially circumspect persons. There are an endless variety of specific things which people may do as responses, triggered by reminders of some great trauma of the past.

They are not random. They are full of meaning and done with precision, on purpose. As problems, they can now be solved.

The Story of Checkers

There is a book by Robin White called *Be Not Afraid* (Dial, 1972). In it is an autobiographical account of the life history of a family with an epileptic son, Checkers. It is a moving story, terrible and horrendous, of the many personal and social problems in a dwindling spiral right to the death.

Here is Mr. White's "explanation" for Checkers' problem: a combination of three "physical" causes, as follows:

1) Prenatal stress, while the brain is forming. He refers to Dr. Solco Tromp's thesis that if there is excessive heat six months before birth, while the cerebral cortex is being organized, there can be decreased protein intake in the fetus, causing brain damage.

2) Birth injury with distortions of baby's head. He cites that the incidence of epilepsy is higher among the firstborns, who may suffer more birth injury by "breaking the way."

3) A critical injury during early years. He theorizes that early adverse conditions (such as the above) must be followed by a critical subsequent injury (such as accidents, head blows, high fevers, strokes, etc.) between birth and age six during which the brain triples in size.

The labored effort to find a rational "cause" is commendable, but he ignores clue after clue in his book which point clearly to the more underlying *real* reasons. For instance:

1) When withdrawn from school, where he was taunted, teased, and bullied, he had few or no seizures.

2) At different times when drugs were reduced or withdrawn—it made no difference.

3) When entered into a psychiatric hospital for study and treatment—away from family and school—he had no seizures.

4) An aversion to watching puppies being born, including one which almost died, and which was massaged back to life.

5) On a world tour, little trouble except a seizure following (37 hours?) a visit to a maternity home.

6) Nearly drowned in the ocean (effort to return to womb?).

All these clues point *not* to "brain damage" but to the existence of triggering *reminders* of a critical event, probably at birth, which posed a survival threat, to which he was responding, time after time. On several occasions when he was away from school and its stress-triggers, he went seizure-free for extended periods of time. As soon as he was sent back to school—again, seizure after seizure. There were many such clues that should have been obvious except for the total *belief* in "brain damage."

What to Do to Solve Your Case

In Chapter 16 will be found the full step-by-step series of instructions for the whole process of problem-solving in the adventure of self-liberation. This includes (and therefore won't be repeated here) such steps as:

1) exploring the desire to change,
2) recognizing your full self-responsibility for your state of affairs and hence its changeability,
3) the diary-keeping which is a fundamental procedure,
4) the ABC's of how to analyze to find the stress-triggers,
5) entering an overall program of deconditioning, and
6) associating with like-minded fellow-seekers in the Great Adventure.

In this chapter certain additional details need to be mentioned because they are specific for epilepsy:

The diary-keeping, problem-solving phase of a search for specific stress-triggers should cover at least a half-dozen seizures or psychomotor responses of whatever kind. Once or twice may not be enough because, to avoid contact with what may be distasteful or painful, one can easily misuse the system. It is quite possible to look at the wrong times and places, find some "good reasons" instead of the *real* reasons, and hence "disprove" the system, saying that it doesn't work. (Read what is said later about fear-avoidance reasons.) It is necessary to observe a pattern that clearly repeats before pinpointing the specific stress-trigger.

The Interim "Bonus Period" of Freedom

There is an interim period in which you are still keeping a detailed diary, you are still "epileptic" and are just beginning to get into the problem-solving stage, but *now* you can begin to go out socially on dates, meetings, concerts, shopping trips, or whatever, *knowing* you will *not* have a seizure! How?

Let's say you would like to go out Saturday evening. It's Saturday morning. Using your calculator, check back each multiple of 12.5 hours (give or take a half hour or so) and *see*, with the help of your diary, whether there were any acute trigger incidents of the kind which you have discovered precipitates the seizures.

Saturday morning, nothing disturbing? Friday evening, Friday morning (8 A.M. to noon—37 hours preceding the whole Saturday evening), nothing? Thursday evening, Thursday morning. Wednesday evening especially (that's the three-day cycle), everything peaceful? No problems? Now Tuesday morning, and finally Sunday evening (that's the six days).

Now, if it all looks clear, with none of the specific trigger-incidents happening at those times, then you *can* go out Saturday evening with the assurance that you will *not* have an unhappy incident. Relax, and have fun!

It is assured that epileptic seizures do *not* "just happen" at any time; they only happen at logically specific times which are predictable, and they depend on events which trigger memories of a survival threat in early experience.

Thus, if you are an "epileptic," by this process you can go out socially, with assurance, even *while* you are working on the deeper problem, and even *before* you have it solved.

That in itself is a revolutionary breakthrough!

A Semifinal Note

Epilepsy is not a disability which you may *have,* but an ability to respond to what seems to be a survival threat, and in a certain manner which you decided was the only way to survive—then. Seizures are something you *do* rather than something you *have.* See the difference? The difference is between being incurable and curable. Now you can pinpoint the specific threat and change the response with that insight.

The convulsion (if that's your style) is like the ultimate retreat inward, away from the threat, which seems overwhelming, and in which your situation may appear hopeless, and you seem helpless.

That hopeless/helpless feeling has been reinforced by the medical belief that epilepsy is due to a permanent brain damage. Actually, the body tissues are fantastically resilient and can bounce back from enormous damage. Such trauma as prenatal stresses, difficult births with forceps and head distortions, or even severe injuries at a later date—*none* of these need be causes for erratic activities far into the distant future! A scar tissue doesn't *do* anything.

Most Epileptics Are Cowards—for Good Reasons!

If you find yourself resisting doing the diary and pinpointing the trigger-source, it's because making contact with the central problem that needs to be solved is frightening. Why? Because back underneath that pattern of seizure responses was a life-or-death decision. The response worked. You survived. And that's life. Now, to confront it, and change to something else, may seem like death!

Imagine being chased by a man-eating tiger. It looks frightening. It threatens. It's scary. But you can stop running away from it, turn around, and put your fist right through it—to discover that it was a paper tiger all along!

This is another way of saying that your fears are really

groundless in the here-and-now. You don't really have to be so afraid; the problem does have a solution; you really can quit the problem; you can be different, you can change, without dying. You're afraid of a paper tiger!

So get with it! Start your diary. It's your first step to freedom. There's a marvelous and wonderful world out there—just waiting for you!

7 The Final Secret of Cancer

NEARLY one-fifth of all deaths in the United States, touching two out of three families, are of cancer. In spite of billions of dollars of effort over several decades it seems that we have made little real progress. With the exception of skin cancers, a diagnosis of cancer is tantamount to a death warrant despite all propaganda to the contrary.

If it is true that we are in command of our body cells as shown by our 2,000 Psycles case histories, what does this say about cancer? Could this dread disease also show motivation? Is cancer just another way in which we self-destruct on purpose?

Because most of us have lost friends or relatives to this horror, it takes a special brand of courage (or foolhardiness) to suggest that these "victims" may have actually been committing suicide, and chosen this way out.

"What?" you might say, "how could anyone *want* to die of cancer, with all that terrible pain and suffering? You must be out of your mind!"

And the answer is that we have been looking at the wrong times and places for the reasons for much of our suffering.

Let's look at some of the "out there" theories about its genesis. They include viruses, auto-immune systems going haywire, bad genes, environmental chemicals, things we eat or do not eat, chlorine in the water, lack of vitamins, embryological throwbacks, and so on, including mutations.

They all ignore the person who is in command of that body.

The Psychological Clues Are Everywhere!

Why must we be so blind to the subconscious motives? Why all the pretentious lies? A wall of hypocrisy and dishonesty usually surround the dying patient, as few dare to express the unspeakable questions: "What's really troubling you? Is there some truth you can't face that is worse than death?"

Actually, there are scores of clues that cancer is as psychogenic as everything else:

1) the sudden healings which are called "spontaneous remissions" and which happen when people change their minds; the many cancers that have been "faith healed" by every which kind of healer or circumstance, whether at religious shrines such as Lourdes in France, or from the humblest of prayer groups.
2) the lack of plausible causes; the stretching of "incubation periods" out 30 to 50 years in order to find "causes" such as carcinogens (vinyl chlorides, styrenes, asbestos, benzopyrenes, PCB's, etc.) which can be in our tissues for many years without causing any effects.
3) cancer cells are being produced and destroyed in the body all the time; nearly everyone has microscopic areas of cancer in various organs after reaching adulthood; nearly 40 percent of males who die of other causes have tiny cancerous areas in their prostate glands; others have tiny tumors in the kidney or stomach; that cancer cells, after generations of experimental tumors, can still revert to normal cells.
4) nearly every type of cancer rate is influenced by marital status; higher in divorced than married persons; higher if no siblings or if an early death of a parent; higher in widowed, divorced, and single persons; 60–70 percent higher rate of ovarian tumors among never married females (whether black or white).

5) many studies show a relationship between cancer cases and some deviation from normal emotional expression: patients who express their negative emotions more survive longer than the suppressives; Galen, the father of medicine, wrote that melancholy women were more likely to get cancer than sanguine ones; LeShan's study in the 1950s linked cancer with a high incidence of loss and hopelessness; William Green in the 1960s found a preponderance of loss preceding leukemias and lymphomas; other studies linked cancer with an inability to express emotions, or a tendency to suppress unpleasant emotions, or an ambivalence in reactions to figure drawings, or to repressed anger.
6) cancer is linked with sexual problems: breast cancer is the commonest cause of cancer deaths among women; there are 200,000 new cases of breast and genital cancer in women, killing 57,000 per year; uterine cancer is higher among prostitutes and promiscuous women; there are four times less cancer in women whose sexual partners have had vasectomies; wives of men with prostate cancer are eleven times more likely to get breast and genital cancers.
7) there is a negative correlation (i.e., reciprocal relationship) between cancer and background radiation, the mentally retarded, and among schizophrenics (who are virtually immune to it); a negative correlation exists between endodermal cancers (of the intestinal tract) and respiratory allergies such as asthma, (i.e., a reciprocal relation between skin/boundary activations and digestive functions).
8) some tumors regress after chemoterapy when the quantity used was far less than needed to kill the tumor; some skin melanomas appear and disappear simultaneously in different skin areas, or remain static for long periods of time; there are antibodies which kill tumors in test tubes but not in the body.
9) cancers do respond to psychotherapy; as demonstrated by Dr. Carl Simonton, M. D., the beneficial effects of chemotherapy can be augmented by psychotherapeutic programs in which the patients with the most positive attitudes and who practice a visualization of antibodies

"eating" up their cancer have significantly more remissions and extension of life.

An excellent example of the psychological control of cancer was reported by Bruno Klopfer in "Psychological Variables in Human Cancer" in *Science* 200:1369.

"A man with advanced lymphosarcoma was included in an experimental study of the since-discredited drug Krebiozen. After one administration, his tumors disappeared. When reports came out that the drug was ineffective, he again became bedridden. His physician, in a last-ditch effort to save him, told him not to believe what he read and treated him with a 'double-strength' Krebiozen—actually an injection of water. The man again experienced rapid remission. Then the AMA and FDA pronounced the drug worthless. The man died within a few days."

But listen to what some die-hard medical proponents of headless bodies have to say: "Jimmie Holland, chief of Sloan-Kettering's psychiatry service and a woman who knows about research in the psychology-cancer area, thinks things have already gotten out of hand. She says there are *no scientific data linking psychological factors and cancer*, either in predisposing a person to the disease, or in determining how well a cancer patient will respond to treatment. She believes that the Simontons, therefore, are perpetrating a 'cruel hoax—worse than Laetrile.' " (*Science* 200:1369)

You see what we're up against?

The Amazing Warts

In medical journals as far back as decades ago, doctors have reported that a surefire way of treating warts among children is to buy them for a penny! (It may take a dollar now.) The doctor looks at the wart and with full seriousness tells the child, "I'm going to buy that wart from you. Here's a penny. Now, that wart belongs to me, and within a few days it will leave you and come to me. OK?"

The child leaves with the penny and, within a week or so, the wart will actually regress and vanish. Something happens in the child's mind, something that had to do with creating the wart. When the impulse to generate it changed, the wart-effort ceased, and the process of homeostasis restored the normal structure.

Now, a wart is in many respects similar to a cancer It is an excessive proliferation of cells in the skin. Cells multiply and multiply, causing an outburst of structure which is not wanted consciously. But subconsciously, there seems to be some impulse—from the mind presumably—to cause the cellular proliferation. Change the mind, and the impulse and the proliferation cease, so that the normal condition is automatically restored.

This is just one example among the many which indicate that motives, ideas, and impulses underlie cellular changes. Pay the child a penny, and the tumor (which is what a wart is) goes away!

The Genetic Finger-Pointing

Throughout medicine there is the unconscious cover-up of psychological reasons for our troubles. A great effort goes into pinning the blame on something "out there" so that we don't have to feel responsible for them and be induced into contact with the painful roots.

A favorite dodge is to blame things on heredity.

In one family four children had died of leukemia. The counselor was about to leave after three hours of "getting nowhere" in seeking hereditary causes. Suddenly, the old grandmother yelled, "We told them not to get married. They're second cousins, and nothing good would come of it!"

According to the National Genetics Foundation, this was an example where genetic counseling should have helped. But is their assumption correct? After all, there are thousands of second-cousin marriages without *any* leukemia.

We believe that if we remove our psychosomatic blinders for a moment, the data should be, on the contrary, just another clue that cancer can be psychologically motivated. In this case the young couple in love were violating a deep belief of their family culture and thereby committing a grievous sin. As a result they were living in mortal terror. And the hypnotic command "Nothing good would come of it!" screamed long ago was guiding their production of children, one after another, who were "no good" and rotten with cancer.

People tend to prove their beliefs in many ways. In this

situation everybody involved agreed that the "sins of the parents be visited upon their children." Each child, therefore, dutifully distorted his inner structure, generating leukemia, and dutifully died.

In another case ostensibly pointing toward genetics, it was reported (*Science News* 1/27/79) that in one family there were three children with three different kinds of cancer.

A study of the family history from all available records showed that there were at least 16 cancers in four generations. To the geneticists this seemed to prove heredity. To us, knowing Psycles, it merely points to kids identifying with both parents and grandparents and reproducing their problems.

We predict that an analysis of the personalities of the children (who they looked and acted like) would show a fascinating lineage of identities through the four generations of family structure. One child, for instance, may be identified with a father who did not have cancer, but who identified with an uncle who did. Another child might have identified with a mother, also noncancerous, but derived the concept/idea/signal/image from a grandmother with whom the mother was identified, and so forth.

What About All Those Carcinogens?

If we recall that various chemicals including asbestos can remain in our tissues for decades without effect, why should we suddenly say *they* are the *cause* when, forty years later, we start becoming cancerous?

The fact is that there are many bad chemicals which we absolutely do not need in our bodies. On the one hand there are many corporations, motivated by greed for the highest profits, who continue to spew out bad chemicals year after year without much regard for the human bodies ingesting them. On the other hand, there are great numbers of people who continue to work, year after year, in coal mines, for instance, getting "black lung," or breathing arsenic in the Tacoma smelter, or taking in Allied Chemical's deadly kepone—when they *could* pack up and go somewhere else and work in clean air. Are they all so innocent of their self-destruction? Are they not suicidal in that they all *know* they are damaging their bodies?

What is the answer?

1) There are bad chemicals that should be excluded from bodies.

2) Corporations should quit polluting bodies (i.e., forced to automate such processes and decontaminate their wastes).

3) *Very importantly:* We must quit inducing people to destroy their bodies by a propaganda which makes almost everything "carcinogenic." The *news* (with its false assumptions) is more carcinogenic than the chemicals!

Consider asbestos: millions of persons have breathed it into their lungs. If the propaganda generates panic, assigns it as a "cause" of cancer and promotes the hope of monetary rewards in indemnities—then a great number of people will be *induced* to generate lung cancers to secure that measure of pity for a "victim" as well as the security of being taken care of. If, on the contrary, we make it clear that asbestos may be an irritant, but that the body can take care of it, and it may exist in the body forever without *causing* cancer, and it's up to you whether you wish to live a healthy life or make cancer out of it—*then* many fewer people will go ahead and generate lung cancers as a result.

Unfortunately, most of the "authorities" in the field are dead-set against the psychosomatic realities we know with Psycles.

The Defeat of Senator Taft

Senator Robert Taft of Ohio died of cancer in 1953.

If it were not for the new insight deriving from Psycles, we would view his death as a courageous fight against an unwanted disease. An innocent victim, struggling for his life, wanting to live, nevertheless succumbed to forces beyond his control.

With this new data, however, and the peculiar set of time-lags which we have discovered between stimulus (the reason why) and the response (cancer), we can now see him as having made a decision to end his life. And with this set of predictable time-lags between trauma and symptoms, we now know where to look for the reasons.

And it all suggests that at least *his* cancer was psychosomatic. That is to say, he was generating it on purpose. He was motivated, subconsciously, to do it because he wished to die.

Let's look at the history of his case, as revealed in an article by Jahn and June Robbins in *This Week* magazine of January 17, 1954.

You may remember that Senator Taft was for a long time "Mr. Republican." After the GOP convention chose the war hero General Eisenhower as their standard-bearer, Taft was invited down to Warm Springs, Georgia, the third week in April of 1953 to talk things over.

Taft was a three-timer loser for the presidency. This confrontation was convincing proof that he was through. Ike was the new boss. The reason for his whole political life was finished.

Without a detailed diary we can't pinpoint the exact moment when he made his decision to die. But the account goes that on Sunday, April 19, with the conference over, they both hurried out to a golf game. At the seventh hole Taft clapped his hand to his hip and complained of pain and stiffness. They finished the round with Ike winning by several strokes. Taft went home not only hurting in his hip, but feeling a shortness of breath and weak in the knees. The symptoms continued with constant fever and anemia, and was soon diagnosed as cancer.

From there it was straight downhill to his death three months later. From the events and timing, I conclude that it began in that hip. His game was up. Subconsciously he made the decision to die, how to die, and began the process of modifying his tissues by proliferating the cells to terminate his life in the manner of a "heroic struggle" on July 31, 1953.

You can choose to believe that "death" is a terrible *thing* that comes to get people against their will. Or you can begin to open your eyes to the wondrous world that really exists. In Chapter 16 we shall see how you can make your life a whole new adventure!

Ramon and His Sexual Identity Problem

In the March issue of *Nursing '77* there is a detailed case history of a young man named Ramon who died of cancer.

When he arrived at the hospital, at the age of 26, he had an excellent position with a large company. He was strong and healthy, charming and friendly, handsome and impeccably groomed. His black hair and full beard were carefully trimmed. He wore flashy scarves and expensive scents. He was a macho Don Juan, separated from his wife, with whom he had a two-year-old son, and was living with a girlfriend.

From the above we would not quickly conclude that here was a person committing suicide. For we are saying that essentially all our cells are under our control and that we are fully responsible for their modifications or distortions, up to and including death. Were there some less obvious clues which might show motivation for a cancerous suicide?

At the age of ten he underwent surgery, a right orchioplexy, to correct for an undescended testicle. That was the first clue to a sexual identity problem. In February 1976, when he reported to the hospital, he already had massive carcinoma of the testicle with embryonal components, an abdominal mass, low back pains, and breast tenderness. His testicular cancer had been developing for several months.

Tests showed an abnormally high level of sex hormones, suggesting a hormone-secreting tumor. With a series of surgeries and chemotherapies, his case was all downhill. Metasteses spread to both lungs. He lost most of his hair and much of his beard. Metastatic lesions developed in his brain. He became blind, had headaches, frequent seizures, partial paralysis, and finally died, exactly six months later, in August.

Why suicide, if this be suicide?

From time to time, Ramon said his illness was the result of sexual relations with women. From all the above clues we can surmise a person caught in the double-bind of a deep homosexual conflict. Identified with a woman (probably mother?), he was yet forced socially into the role of an aggressive male. So while he wore a beard and acted macho, he was nevertheless subconsciously destroying his testicles and even attempting to grow breasts and a womb (the breast tenderness and the abdominal mass). His concealed feminine identity was also manifest in his fastidious grooming, his flashy scarves, and expensive scents.

Why his deep internal conflict (identity versus biological structure) should be suicidal can only be determined by much additional data and relationships not detailed in this account. If we rest on surface appearances, we can be carried away by the tale of an innocent victim, struggling courageously for life. But if we look underneath the surface, and especially at the first tissues that are distorted and proliferated, then we may begin to gain insight into motivation.

By asking the most appropriate questions, "How are you identified, and why are you trying to kill yourself this way?" the underlying problems and conflicts might have been opened up for solution, to reverse the whole process.

But no one questioned his motives. And Ramon died.

A Spontaneous Remission

What happens when a cancer suddenly quits? The word remission is interesting in that it suggests that you have "paid your dues" and can now be free of the problem. There may be more truth than fiction in that. There are scores to be settled, punishment to be suffered, dues to be paid.

A friend of mine told me about her mother. Some sort of stomach-intestinal cancer had spread beyond treatable stages, and the two sisters were arguing what to do with their mother. They were at loggerheads over whether one of them should drive the mother to Texas to the Hoxsey center and who could better afford the time and expense. Apparently the argument with its pros and cons had gone on for days when, suddenly, the mother said, "A pox on both your houses!" or "To hell with both of you!" (I don't remember which) and proceeded to vomit up a horrendous black mass of putrid material.

Within a day or two she was well. The cancer was gone. It seemed as though she could "stomach just so much" before she decided she had had it! This of course does not touch on the deeper roots of identities and motives, but it does show something of the psychological aspects underlying at least one so-called "spontaneous remission." Behind every one of them you can be sure there was a deep turn-about in attitudes, feelings, and relationships.

How Connie Was Misnamed

Among the many clues to psychological realities that are just underneath the surface of our troubles, are the problems that can go with improper naming. Names should be carefully chosen with the knowledge that people often live up to them (albeit unconsciously). Daughters shouldn't be named after fathers, or sons after their mothers, for instance, unless you wish to foul up sexual identities.

In Connie's case it brought death of cancer.

Her father's name was Lucius Constant, and he was called Con. Their second daughter was named Constance, or Connie. Until she was married and became pregnant for the first time, there was no problem. She was a gorgeous blonde, socially prominent at school, a high-spirited and fun-loving date. But troubles started after she delivered her first child.

Deep within her subconscious there was a conflict between the name and its father-identity, and the female biological structure which she had. So, step by step, she began destroying those very organs which contradicted the secret self-image. First an ovary, then another one. Then a hysterectomy and the removal of her uterus. Finally, first one breast and then the second, with a radical mastectomy. Who would ever suspect, except me, her brother, that she was deliberately (however subconsciously) destroying, one by one, her female organs? By the time I got to her, with deep heart-to-heart talks about hidden identities, it was too late. Her suicidal self-destruct was a total commitment, and I couldn't break through that wall of deceit, reinforced by a lifetime of roles and successes.

Exploring Bob's Cancer in India

Shortly after I registered at the small hotel in Bangalore, I found myself talking with Bob, age about 28. He said he had come to India because he was dying of cancer, all his other options had failed, and he was hoping for a miracle from the Indian avatar Sai Baba.

"Ah!" I said, "I'm a biologist, and I have always wanted a good case of cancer to see if I could penetrate the fundamental reasons why you (for instance) would be trying to kill yourself that way!"

He said, "Be my guest."

Thus it was that during my extended visit in the area during

1974 I had many lengthy conversations with him, tracing the history of his disease, and in fact his whole life, searching for clues to its psychological origins. He was totally cooperative.

"Between your mother and father, who did you take after?" I asked, after suitable preliminaries.

"My old man," he replied. He owned a dry-cleaning establishment in Flushing, New York. Little by little I gained the picture of a man who was full of self-hate, hostile, paranoid, tense, dictatorial to his wife, and hard on his kids. I sensed that the business he owned satisfied a deeply symbolic function, equivalent to the Macbeth "out, out, damn spot!" in a continual if not futile effort to flush out some deep feelings of evil-dirt-guilt. The fact that the business was in Flushing was in my opinion not coincidental.

What was Bob's cancer? Most significantly, it permeated his lymph system, the "sewer" of the body, and was called Hodgkin's disease. And when did it all start? Bob was in Madrid at the time, caught in a romantic triangle. On the one hand was a girlfriend who had pursued him for years and wanted to marry him. On the other hand was a young man with a flashy sports car who wanted Bob to drive with him to Paris. In choosing the latter, he was making a choice favoring the homosexual aspect of an identity conflict that had been brewing for years. Within days (one or two Psycles) he generated his first symptoms of Hodgkin's disease.

During the several subsequent years he had radiation in London, for a temporary reprieve, then chemotherapy and psychotherapy in New York for another reprieve. Relapsing again, he had surgical removal of his spleen in London, which helped for a while. After another relapse, he tried acupuncture in London. This seemed helpful for a while but was followed by another relapse.

It was then, after having exhausted all his options in the West, that he read about Sai Baba and his "miraculous" healings, and decided to go to India. There, after several months of additional suffering, his final "cure" is a tale of magic that could only come out of India.

Living at Sai Baba's ashram under rather spartan conditions, he reached a point where he was weak and emaciated, could eat very little, and was in constant pain. Word came from Sai Baba that he should go back to Bangalore and enter a comfortable hotel.

Now comes the incredible events associated with his "cure."

He was walking toward a medical clinic where he was to have a biopsy of his lymph nodes, when there was a commotion up ahead. People were rushing toward a construction project on the right, from whence he could hear loud cries of anguish. He reached the area, and stepped up to peer over a wall, and took in the following situation: the father was a carpenter, his wife was helping him; they had left several children unattended. Their son Ganesh, one and a half years old, had drowned in a small fishpond. The father had the boy by the ankles, trying to shake the water from his lungs. The mother was screaming hysterically over the loss of her son. When the father set the boy down on the ground, the body was stiff with rigor mortis.

Into Bob's mind flashed the thoughts: "Sai Baba, don't you do this to me! I don't want to be any part of that kid's death! If you are what they say you are—bring him back to life!"

In the next instant, the parents noticed a small flicker of Ganesh's eyelids. They scooped him up and rushed out of the compound, out into the street and up the boulevard toward the hospital with a large crowd following.

At the clinic, Bob followed the parents into the operating room where he poured some of Sai Baba's *vibutthi* (sacred ashes) over the boy's chest and head. The female doctor checked and found a pulse, and ordered plastic tubes to pump water and mucus from the boy's lungs, after which Ganesh was bawling loudly, very much alive!

Shortly thereafter, Bob's biopsy turned out negative, and he was invited to return to New York. When Bob told me about that experience, I said, "Strange about that. Did you, by any chance, almost drown when you were about the same age?"

"Not that I know of."

"Because, somehow, it may be uniquely significant, as though Sai Baba, with his subconscious omniscience, were reaching deeply into your past, to turn death into life at some deep core of your early experience, as part of a deconditioning process, to unlock some binding association underlying your cancer."

Bob returned to New York. Strangely enough, the day before his return, his father suddenly died, as though he were deeply locked into the son's cancer and the "cure" was something he couldn't face.

Several weeks later I received a letter from him. "Guess what! I asked my mother about any possible near-drowning, and it turns out that I *did* almost drown when I was about one and a half years old, and got rescued in spite of myself. You're right again! I was crawling around a swimming pool in a resort bungalow colony with my mother and aunt in the immediate area. My mother heard the splash and (being out of the pool and in a lounge chair or something) *froze*. My aunt, who was in the pool in a rubber tire tube, was hopelessly entangled in the tube and could not get out to rescue the little fellow. A third party, a German woman, jumped in the pool and got me out. And here I am today!"

The Indian episode was in September 1974. Was the "cure" permanent? On March 28, 1977, he called me from New York to say that he was on chemotherapy again, having relapsed, but was going to India again in December. I received a card from him on March 22, 1978: he was now back in New York "in best shape yet!" This time Sai Baba said to him, "No more problems." And Bob says, "I won't argue."

Does not all of this say something about the psychological aspects of cancer which is so vehemently denied by the Sloan-Kettering psychiatrist? Such blinders!

8 The Social Problem of Being Sane or Insane

WE have been saying that much, if not all, that we do is on purpose. Beginning with "accidents" and then going on to various illnesses and diseases, we have shown, through the use of Psycles, the reasons behind many of them. We are truly masters of our own destinies, bringing about our ups and downs with incredible precision. The same can be said about our behavior.

Strange Behaviors Are Logical!

Who hasn't at some time behaved in a manner which was embarrassing or out of context with their usual personality? Perfectly normal people can suddenly do bizarre things, and then wonder whether their minds are cracking. To provide a much needed relief for these perplexing enigmas of behavior, Psycles comes to the rescue by showing that they are logical, delayed responses to specific stresses.

We have cases of impulsive propitiations, busy panics, clashes with friends and doctors, depressions, nightmares, crazy feelings, and a whole spectrum of the unexpected, which were instantly clarified by contacting the motivating incident at the earlier circadian timing.

When we get "all mixed up," this is on purpose. When we just happen to run into someone, we meant to. When we lose something and can't find it, we really know, subconsciously, exactly where it is all the time, but dramatize the hunting. When someone steals something from us, again we determine what was to be stolen and, deep within, know who and why. When we suddenly get into an argument with a friend, there is an agreement to "conflict" in perfect cooperation for the expression of *both* our needs.

Let's look at a specific example of "insane" behavior to show its motive and solution. It is only one case in many from which we derived our historic conclusions about life and human troubles. One by one we have taken these troubles, unlocked them with the Master-key Psycles, seen something of *why* they happened, and now are putting it all together in this book for a new vision for a new era.

My Crazy Mother

It was in Claremont, California, in April of 1952 when my mother suddenly "went crazy." Fortunately it was not for long, because I just happened to be visiting, saw it, discovered why she did it, and helped her unlock it.

The occasion of my visit was my recent return from Thailand, where I had been Vice Consul in the Bangkok Embassy. My mother was in good health, served on some civic committees, was writing a book, had good friends, some 14 grandchildren, and was not obviously scheduled for a psychotic break.

On that Monday evening at about 8:30, we had finished dinner. She cleared the dishes and was in the kitchen, while I remained seated at the dining table. Hearing some strange sounds from the kitchen, I rose, turned, and faced the shock of my life. She was standing at the sink, fumbling dishes, mumbling incoherently, her eyes staring into space. Confused, disoriented, she was totally out of contact with present time.

I did the socially proper thing, asking what was the matter. She stammered a meaningless series of words, appropriate for one who had "lost her mind" or just gone crazy. Even while a chill wave of fear swept over me, I glanced at the clock, made a quick calculation, and asked, with a finger snap, "What happened yesterday morning? Sunday morning about seven . . . where were you, what were you doing?"

A "miracle" happened. Her aimless, confused motions and her gibberish ceased. Her look of terror softened, and a light returned to her eyes.

"Why, why yesterday morning—early—I took a walk, and for the first time, down the road about two miles, I visited your father's grave. I had avoided it ever since my return from Thailand. . . . "

And almost as suddenly, she was her old self again. She was directed to relax on her bed, breathe deeply, and review the walk with its emotions. Mentally she was to go back to the grave, see it, feel it, recall her thoughts. And that was the end of her psychotic episode. No further troubles.

Now let's see what sort of meaning we can derive from what happened. She went to my father's grave for the first time and, 37 hours later, went crazy. As part of the background to this situation was the fact that she flew off to Thailand without my father's blessing to the northern capital city of Chiengmai, where she set up a unique little hotel, which became rather famous for tourists seeking the unusual out-of-way places. They could sit in her riverfront pavilion and watch the elephants maneuvering teak logs in the river.

It was while she was there that my father died in California, leaving my mother with some deep feelings of guilt over his death. Now home and only two miles from the cemetery, she had nevertheless avoided going to see his grave until that day. Standing at the graveside, that terrible and overwhelming feeling of guilt rose to the surface. Past events could not be canceled. To handle it, she chose one logical way out. Thirty-seven hours later was her act of insanity, a dramatization engineered subconsciously in accordance with one of many models for insane behavior. And all was forgiven.

Why? The law says: "Not guilty, by reason of insanity."

For how many patients in mental institutions today is this their solution for an intolerable problem of guilt? Let's explore the general problem.

The Underlying Fears

The problem of insanity touches on every human being and is one of the earlier things we have to confront when we embark on the adventure of self-liberation. To understand it, we must consider the basics:

1) pain is the great warning signal for survival;
2) we learn by experience and association;
3) to survive we avoid not only pain itself, but that which is *associated* with pain, and also seek out that which furthers life;
4) under certain extreme conditions of a life-or-death stress, we identify deeply with that model/image which was associated with survival; because of the unconscious condition at the time, it is "locked into" the "subconscious."
5) we are progressively "conditioned" thereby, in the quantitative, or cumulative process of becoming normal, social human beings.

Now, as a child at birth we have already experienced pain; we have been tuned into the tensions and mis-emotions of our parents for nine months, and have formed many memory associations thereon.

But because our quantity of conditioning is still relatively small, we are much more aware of unconscious realities than adults, and more apt to reflect spontaneously, as bright and shiny mirrors, their hidden messages and secret identities which we so freely perceive.

Unfortunately, that awareness is much too painful for the adults to tolerate. Therefore they, and the whole society in which we appeared, exerted great pressures to force us to shut it off. They denied and ridiculed what we knew was real, saying we were silly, stupid, or crazy. "Nobody told you that, how do you know?" Ultimately they threatened us with violence if we didn't conform to their conditional ways.

And we were faced with a terrible choice—to agree to identify in some acceptable manner with a shutting-off of our subconscious awareness and thereby to accept what is essentially spiritual death, or die.

To disagree would have meant no food or shelter, an excommunication from society, "accidental" death by suffocation or "sudden infant death syndrome" or any of a hundred other ways that children's lives are terminated.

In order to survive, therefore, often by the age of two and usually by the age of five, we agreed to surrender the cosmic and become deaf, dumb, and blind. We suppressed that divine spark and learned to be liars like everyone else. We retreated into the conditional. We *had* to, to survive. All of us.

As adults, therefore, when we think of insane asylums we are reminded of the time when we *knew* things and father and mother and other big folks *didn't*. We remember the threats and the loneliness, the terror of being rejected and denied and punished—when we were too small to survive alone. We are reminded of what society can do to excommunicate when one doesn't conform, and of the momentous decision we made to limit ourselves and accept spiritual death. And we are afraid.

Fortunately, however, society doesn't really fear the unconditional. It is the antisocial reminders of pain that it fears. It permits the deconditioning for gradual self-liberation, if it is done discreetly.

What, Then, Is Insanity?

From the massive evidence of Psycles case histories, we can conclude that no one *goes* insane, like a ball rolling down a hill. It can't happen *to* you against your will. Each case of insanity is engineered, desired, willed, motivated, and created with skill. We are not "victims" of insanity. Insanity is an *act,* done with precision, for reasons, in accordance with a model.

Why, then, do many people *wish* to manifest un-sane behavior?

Let's not get carried away by the engineered buildup. The real reasons may not lie in the series of events leading up to the "break." Underneath all the careful planning lies a deep identification, and a choice of alternatives. In the one case we just looked at, there appeared to be an alleviation of an intolerable feeling of guilt. But there are many other possible reasons for choosing this type of behavior. Among them are deep identifications whereby the *doing* of un-sane behavior brings some important side benefits, as in the next case.

The Schizophrenic Artist

Attending art school in her mid-twenties was Tina, who confessed to a mental panic because, "recently, I've been hearing voices." She had previously been committed to a mental institution for about three years, but had been out for an equal period of time, functioning fairly normally.

"When's the last time you heard voices?" I asked.

"Yesterday morning about 11." That would be Thursday. "I had a very heavy panic, like a cloud closing in over me and a voice suddenly said 'get a knife and slit your throat!' "

"Gosh! What was happening with you Tuesday evening, about ten?"

"Sound asleep. From six or seven that evening. Same Wednesday."

"OK, let's go back three days, then. What about Monday morning around 10?"

"Now, there's something! A real cute teacher in art school cut me down saying I was wasting paper! That was real mean; I thought I had a good drawing. I felt hurt . . . rejected."

"What sort of guy was he—masculine, or more arty?"

"Beautiful! With lots of style!"

"OK. Between your folks, who did you take after more?"

"Daddy," she said as she described how when he left, and she was only five or six, she began acting and behaving more and more like him. The more she was like her daddy, the more her mother liked it, and liked her. In fact, it transformed their entire relationship; mother and daughter became pals.

"Your old man ever had any mental troubles?"

"Strange you should ask. I didn't know about it until much later, when I was told by my mother. He was pretty schizzy, and had to be put away."

"Aha! There's your problem. When you identified with your father to get *with* your mother, you didn't bargain for his being crazy, too. But here you are, tied in to that identity, and not too keen on giving it up, either, because it's like you'd lose all that love and contact with mother. But some day, hopefully, you can get back into that decision, and unlock it."

Once we exposed the pattern of secret identity, her case and problem became quite clear.

She had very good reasons for identifying with her father: at age five, before he left, she was in serious trouble with her mother. But after he left, the identity saved her life, bringing the two together as pals. At this point she had to confront the question of whether she could still relate to her mother without that identity. Would it mean the loss of a loving relationship? And how important was the insanity-act feature to her mother's deep needs? After all, her mother had selected for a husband a person who dramatized this property.

There were also other ways in which she used this mental problem as a weapon to get certain things. She had a way of threatening to go crazy if people didn't treat her rightly. These were subordinate problems to the one of relating to her mother, but needed to be thoroughly worked out as well.

The Search for "Causes" Elsewhere

Hundreds of millions of dollars have been wasted on efforts to find some chemical basis for schizophrenia, and every year some researcher shouts "eureka!" pointing to some metabolic chemical in the bodies or brains of crazies which is a little more or less than what is found in the non-crazies. This search goes on for two reasons: 1) psychologically, the problem has always been terribly difficult to solve in the absence of Psycles time-lags; 2) chemically, there have been some strong pluses: tranquilizers are generally helpful, and lithium salts have leveled off manic-depressives. It was convenient therefore to conclude that mental problems are "chemical," and some magic bullet could be found to end them.

From Chapter 3, however, we found that even though a problem has psychogenic origins, there are many chemicals as well as other physical forces which can oppose the conditional *responses* at various levels to help turn them off.

Until the magic bullet is found, therefore, the problem is still acute, and this chapter must deal with the essential psychogenesis of mental problems and how they may sometimes be solved with Psycles.

It is sometimes convenient for diagnostic purposes to assign the cause of a case of insanity to a brain tumor. From our Psycles data, however, we would say this was putting the cart before the horse. In other words, look first for a serious problem-conflict within the patient's mind. He/she may be transacting for a state of insanity for certain reasons, and generating the tumor as part of an effort-system within the brain.

Start with the assumption that we really get what we want. Don't be afraid to ask why some painful condition might be desired. This way we can bypass the trap of the many dubious "reasons why," such as genetics, which derive from efforts to avoid self-responsibility. This will bring us much sooner into contact with what's really happening.

On Going Crazy

In early 1976 the senior editor of *Time,* Otto Friedrich, published a book called *Going Crazy,* in which he said that most of us have experienced twinges of craziness and that it wasn't an either-or kind of thing, but more like a gray area in which we may all occasionally enter for moments or more extended periods of time. We should quit feeling that it's us or them, sane or insane, as two distinct species.

So far correct. But when asked about causes and cures and what-to-do's, he said that it was probably a chemical derangement to be best treated chemically. False. (Except that every time we raise our little finger a million chemical changes occur.) The fact that a feeling of craziness can be turned off or cured by chemicals does not mean that its cause is chemical. (See Chapter 3.)

He described how he once came to a street corner and suddenly felt completely incapable of knowing what to do to cross the street. It was like a blackout of all previous experience on crossing streets up to that moment. He felt confused and bewildered.

Without Psycles, his only recourse was to go to a hospital, see a psychiatrist, do some talking, take some tranquilizers, and hope that the trouble didn't persist. With this avenue of action, however, he had the gnawing suspicion that "something wrong" was happening to his mind or brain, like "going crazy." Was he suffering from some chemical lack?

But with Psycles the picture is entirely different. At that moment, or some time later, he could scan what happened 37 hours or three days plus a few hours earlier, and discover the *reasons why* he did this event on purpose. He could discover the meaning and motivation for feeling this way.

He would have no haunting worry whether something was "going wrong" in his head—chemically, or any other way. He was merely doing something he needed to do, on purpose, to "balance the books" for some event which happened earlier. And seeing the connection would either be its permanent "cure" or provide insight into a problem which needed attention and could be solved.

The Disoriented Carpenter

On a Monday morning at about 9:30, a carpenter was busy putting up siding on a new house when he suddenly became dizzy and disoriented. Hoping to dispel the mental confusion, he stopped work for a cup of coffee, and then begged off for the rest of the day.

The source of this problem was pinpointed with Psycles. On Saturday evening, 37 hours earlier, he went to see a movie called *The Ruling Class*. In this film, Peter O'Toole played the role of a British lord who was also a psychotic Jesus. Returning from an insane asylum to take over the baronial estate, he brought with him a huge cross, on which he hung periodically.

This modern-day carpenter, with a secret Jesus self-image, came "unglued" 37 hours after the film, but his confusion terminated just as quickly when he made the connection with the movie.

The Safety Inspector, About to Lose His Mind

Jittery and disoriented, a safety inspector felt like he was about to lose his mind. His panic nevertheless subsided almost instantly when he connected with a stress-trigger incident 37 hours before. He had received a phone call, telling him how an important person whom he admired, almost as a heroic model, had been hospitalized and was being given shock therapy.

A Sudden Depression

Alice was 55, long divorced, working as a secretary in an insurance office and barely making ends meet. One Wednesday afternoon at about 4:30 she was returning home on the bus, when she suddenly felt a deep depression. There was nothing within the immediate hour or two to account for it. It had just come over her: life was unbearably hopeless, her future bleak and despairing. Why the sudden depression?

Backtracking 37 hours brought nothing. She was sound asleep at 3:00–4:00 A.M. But doubling back to three days and several hours brought the answer. Sunday afternoon between 2:00 and 3:00 she accompanied a friend on a visit to a beautiful new set of condominiums. While walking through these apartments, she remembered feeling a deep but repressed bitterness that here she was, age 55, still working and struggling for a living, and unable to afford a decent and comfortable place to live. She felt deprived of the pleasures which so many others were enjoying, and thought how unfair her economic situation was, with little prospect for changing it at this late date in her life.

That total vision of her life as triggered by the visit to stylish quarters—her essential failure, her inability to afford comfort—was the immediate cause of the depression, three days later.

Making contact with the trigger incident brought immediate relief just in knowing why she had the depression. But more importantly, it provided contact with the inner conflict and the feelings about her nonachievement, and initiated a more frank appraisal of her chances and options in life. In a seminar in which this sequence of events was discussed, she was bombarded with an array of adventurous possibilities in life which were quite within her grasp.

A Suicidal Impulse

Ruth was returning from a steno refresher course Monday afternoon about 1:30 when, right on the sidewalk, she broke down crying, and between 2:00 and 3:00 was hysterically threatening suicide. A close friend spent the time diverting her with talk, food, music, and motion. Why this sudden impulse?

At 37 hours before, she remembered a restless night and feeling depressed, but no untoward events. Doubling back to three days provided the answer. She was in the middle of a divorce. She was Catholic; this was her first and only marriage, and she had a 15-year-old son. Friday afternoon, while she was conveniently away at the post office, her husband came and "abducted" her son. Three days later, the suicidal impulse.

What elements went into this? Her lifetime Catholic upbringing dictated a mother's responsibility for the son. But the son was incorrigibly poor in school and a serious drag on her efforts to start a new life for herself. With the suggestion that she might really wish freedom from that responsibility, that she really didn't need to feel guilty about relinquishing him, that he was a serious problem, that he preferred to be with his father anyway, that the "abduction" was confluential (cooperative), and that she was intentionally away because she knew subconsciously that her husband was coming to get him—an enormous weight lifted from her shoulders.

Each person is unique. What triggers one person will not necessarily trigger another. And what we do by way of response reflects our own unique backgrounds. Only by exploring the specific cases do we see how Psycles is manifest, and how it reveals hidden aspects of personality and motivation.

On Losing Things

In Chapter 2 we told the story about the lost car keys, and how it symbolized the feeling of being shut out of his wife's "world." From our various cases we must assume that all "losses" are spurious in the sense that we really know, subconsciously, where the lost object is. We are merely dramatizing the *feeling* of having lost something important.

Now let's look at another case, a lost wallet, to see what it can teach us.

On Tuesday afternoon between 4:30 and 6:00, George was feeling extremely upset because his girlfriend triggered old memories of wifely infidelities. He felt the loss of a special magic in their relationship because of a complicated arrangement, involving the use of her bed by another man, with whom she insisted she had no romantic interest.

Friday afternoon (three days later) George returned from a shopping trip about 5:45, changed his clothes, and left his wallet in the pants he hung up. Now it wasn't until Saturday morning about 10:00 when he missed his wallet, searched high and low, went into a panic, revisited the shop, bank, and parking locations, and even went to the motor vehicles department—which was closed. After six hours of worry and upset, he "found" the wallet in the other pants, at 4:00 P.M.

It's of interest to note that the act of losing the wallet was at the circadian timing, but the loss *feelings* drama didn't occur until the next day. Why 10:00 A.M. Saturday? Was something happening 9:00 A.M. Wednesday between his girlfriend and the other man, after he had spent the night in her bed (she slept in the living room, according to her account)? Or were his suspicions and anxieties merely an expression of earlier traumas in his own life—projected onto the present scene?

The Missed Plane

A lawyer flying from Los Angeles to Seattle, had a stopover in Portland. He was informed that there would be a 40-minute stop, and the plane would depart for Seattle at 10:40 P.M. He decided to stretch his legs, pick up a snack, and then returned to the plane with ten minutes to spare, only to find that it had already departed, leaving him stranded in Portland.

He stormed against the airline, threatened suits, and demanded the next flight on any plane to Seattle as he had friends waiting at that airport. With a half-hour delay, he continued his flight.

If we assume that this whole event was engineered on purpose, with all the appropriate subconscious knowledge to make it come out correctly as it did, then the question is: for what reason would he wish to have his flight interrupted, to be abandoned in Portland, and to feel rage against the plane and its airline?

The answer lay clearly and precisely in an incident which happened that morning. At about 10:00 A.M. (one circadian half-wave instead of the usual 37-hour lag), he was in the middle of an important legal examination, when the examiner, a female, interrupted him with an infuriating question on procedures. He was so outraged by her action, yet so blocked from responding, that it took him a full half-hour before he could compose himself sufficiently to continue the exam.

A short Psycle later (the first inverse geophysical function) he engineered the situation in which he could vent his anger against a huge female symbol for causing a half-hour interruption in his "flight."

The Rambling Speech

A corporate executive who was an accomplished speaker with years of experience found himself rambling on the subject of death during the course of an important speech.

It unnerved him. Was he slipping mentally? Losing his command of subjects and audiences? The more he thought about it, the more disturbing it was.

Fortunately for his complete self-assurance, the reason for this strange departure from his normal behavior was found and all his doubts erased. What happened was this: three days and two hours before the speech, he was suddenly confronted with the news of a shocking death, which touched him deeply and personally. Then, 37 hours later, a second experience happened in which there was a tragic death apparently due to mismanagement. Now, at speech time, he was responding to a one-two punch, with both events zeroing in on the same time, causing him to veer off from his prepared text, to his great consternation.

The pinpointing of the trigger incidents, with their hidden stress, was like the sunshine breaking through stormclouds, bringing release to a mind that was sorely troubled.

Troubled people cost industry far in excess of a hundred billion dollars per year in unproductive wages, in damage to themselves and their machines, and in product liability and lawsuits.

Knowing the roots of troubled behaviors can often terminate

them. And anything that can deal effectively with solving people's problems will at the same time help to reduce this enormous cost burden to business and industry. That is *besides* bringing greater health and happiness to all concerned.

9 Ah! Death! A New Look at an Old Bugaboo

For ten thousand years and longer we human beings have pondered the puzzle of death. What is it? And what should we do about it?

A thousand different ideas, beliefs, and institutional practices revolve around it. Death has been personified as an evil force, as a power, a thing or entity which is mysterious, fearful, and unknown. It has been labeled as an enemy that is unpredictable, spiteful, or vengeful. As the grim reaper, it waits patiently and then swoops down to seize the unsuspecting victim to carry him away struggling.

The problem of accidental death—the idea that a man can die prematurely, against his will—has been a special thorn in the side of philosophers, resulting in much sidestepping of reality and the inventing of a great many explanations, most of which are nonsense.

For with the revolutionary breakthrough of Psycles, we can state emphatically that death is not accidental, never was accidental, and never will be accidental. By this we mean that death is not by chance, unwanted, unmotivated, or meaningless. And therefore we do not have to go round and around on those stale questions: Is it good or bad? Do we survive it or not? If we don't, be stoic and live the good life. If we do survive it, prepare for heaven to live happily ever after. Or if it's to be hell—tough luck!

Through all this nonsense the greatest single truth has been ignored with a passion, and shunned like poison: Death is *also* self-responsible, motivated and purposeful—appearances notwithstanding.

Technically speaking, no one dies against his will, and all deaths are suicides. Though it has been said by mystics, this profound truth has usually been obscured by verbal poetry, because it is so difficult to accept. In the Bible, Jesus speaks of the lilies of the field, and of sparrows, and how the hairs of your head are numbered—all by way of indicating the micro-precision of events, that nothing is really "accidental" to God. Now, with the Psycles Master-key, and recognition of circadian reaction time-lags, the fact becomes self-evident. What the mystics have been saying is true. It need not be taken on hearsay, or on authority, but is the conclusion to be derived from specific cases and what they did. The underlying time-lags are the same, whether in the biological rhythms of microbes or the circadian timing to be found in accidents and psychosomatic responses to stress.

By now, of course, you know that when we say that death is self-responsible, willed, and technically a suicide, we do not mean that people generally plan to die *consciously*. We are speaking of the *subconscious,* which is very real, and magnetically connected with the rest of life.

One of my friends had a grandmother who was approaching the century mark. All her friends had long since died. Her health was surprisingly good for such an old lady, but she wasn't really living for any good reasons. She was just hanging on, existing.

One day, he sat down with her and had a long talk about life and death, and reminded her of some religious assurances and biblical promises. The next day grandma called the family together, settled all her affairs, and precisely 37 hours after the talk, she quietly and peacefully expired, with a smile on her face. If a person decides to die, and *does* it, is this not "suicide"?

Evidence that Death Is Not Random

One peculiar set of clues pointing to self-responsibility for deaths is found in insurance statistics. There are significant drops in death frequencies during the periods immediately preceding national presidential elections, important national holidays, the Yom Kippur day of atonement for the Jews, and for birthdays.

In the Kunz study, for instance, nearly half of his sample died in the three months *after* their birthdays, whereas less than 10 percent died in the three months *before* their birthdays. The patterns existed regardless of whether the cause was "natural," an illness, or an "accident." People definitely *wait* for their birthdays. For accidents, the rate actually rose to more than half during the quarter following birthdays.

In addition to these proofs that we have at least *some* control over *when* we die, there are also scientific studies focusing on reasons *why*. A five-year study by sociologist David P. Phillips of the University of California at San Diego (*Science* 201:748) showed that following well-publicized accounts of murder-suicides, there is a sharp peak at three days (two Psycles) for both motor vehicle accidents and fatal crashes of private and business aircraft.

With the predictable time-lags of Psycles, we have an entirely new breakthrough to bring light and understanding to the age-old mysteries of death.

Why Michael Rockefeller and Max Baer Died

Back in November of 1961 young Michael Rockefeller was on an anthropological expedition in the jungles of New Guinea. On Friday the news media carried the story of his father and mother's deciding to separate and divorce. On Sunday, he was reported missing, having been last seen drifting out to sea in an area infested with crocodiles. A divorce of father and mother is not sufficient reason for most persons to wish to die, but was it a special blow for Michael?

Max Baer, former heavyweight boxing champion of the world, died suddenly of a heart attack in Hollywood about 9:00 A.M., Saturday, November 21, 1959. On Wednesday evening he had just refereed a nationally televised bout in Phoenix, Arizona, and had come to Hollywood to participate in several TV appearances, to be followed by numerous business dealings in Sacramento.

Going back a Psycle to Thursday, at 9:30 P.M., we find an obviously significant event. A TV special featured prizefighter

Ingemar Johanssen in a 90-minute live dramatization of Ernest Hemingway's story "The Killers." In this very paranoid story, two gunmen close in on an ex-prizefighter for the kill. Identified with Johanssen, was he reminded of some unfinished Mafia deals—to decide that death was the best way out?

Clark Gable and Marilyn Monroe

Clark Gable was in perfect health on Friday, November 4, 1960, the last day of the filming of *Misfits* with Marilyn Monroe. But then look at what happened Psycles later:

Friday	9 A.M.	Last day filming *Misfits* with Marilyn Monroe.
	4 P.M.	Finish, and goodbyes.
Saturday	9 P.M.	(36 hours) Went to bed feeling "sick."
Sunday	4 A.M.	(36 hours) Awakened with chest pains, perspiring, clammy. Heart attack. Rushed to hospital 8 A.M.
Monday	4 P.M.	(36 hours) Relapse, in hospital.

Nine days later, on November 16, 1970, he was dead. Do we not sense the tragedy of the "King" and "Queen" who just couldn't come together? as "misfits" with no way out but death?

The Incredible Kennedy-Lincoln Drama

I hesitate to mention the death of President John F. Kennedy, because so much has been written about those days of Dallas in 1963. However, it is such a grandiose example of identification—Kennedy and Lincoln—and carried out with such extensive cooperation by persons with parallel identifications, that it would be difficult to ignore.

Here is a brief review of some of the identifications:

Lincoln and Kennedy were both concerned with Civil Rights and freeing the slaves.

Lincoln failed as a vice-presidential candidate in 1856; Kennedy failed as a vice-presidential candidate in 1956.
Lincoln took office in 1860, Kennedy in 1960.
Both were slain on a Friday in the presence of their wives.
Both were shot in the back of the head.
Booth shot Lincoln in a theater and ran to a warehouse; Oswald shot Kennedy from a warehouse and ran to a theater.
Both were succeeded by men named Johnson, who were both southern Democrats and ex-senators: Andrew Johnson born 1808, Lyndon Johnson born 1908.
John Wilkes Booth was born in 1839; Oswald was born in 1939.
Both assassins were slain before they could be brought to trial to testify.
Both cases reeked of conspiracy in high places.
Both Presidents' wives lost children by death while residing in the White House.
Lincoln's secretary, named Kennedy, advised him not to go to the theater; Kennedy's secretary, named Lincoln, advised him not to go to Dallas.
A New York policeman, named John Kennedy, frequently expressed fear for Lincoln's life.

The point is *not* all these "strange coincidences." One has to see people making decisions to identify in certain ways and then carrying out these roles. From the Lincolnesque fulfillment, we can infer that at some early age, such as two, John F. Kennedy must have made a deep decision to *be* like Lincoln. We can also assume that the decision was of survival significance in his relationship with his father and mother. Having identified, and finding success in that identity, he then proceeded to carry out that life . . . right to that death.

When we learn to "see" people all around us acting out roles and identifications, we may also recognize the enormous and automatic cooperation which we all get as well. The fact that it is all "subconscious" does not mean that it is not real. The central challenge of life itself is to reverse that process, to dis-identify, to unlock, to open up, to decondition—and hence free ourselves for the incredible life!

Touching a Few Other Deaths

A much less historic example of identification and death is that of a 17-year-old cinema usherette, Joyce Funk, who was standing on the sidewalk whan a car careened over the curb, struck her, and dragged her body 20 feet. This was in Cincinnati, Ohio, March 10, 1955.

What reveals the identification is the fact that three years earlier, her brother, then 17, *also* a theater usher, was struck and killed by a trolley bus while riding his bicycle at a spot hardly two blocks from where Joyce was killed.

To cooperate with her intention to be killed, the driver (grabbing at a door that flew open) had to "accidentally" come up over the curb and onto the sidewalk to do her in. Newspapers are full of accounts of strange and "coincidental" accidents and deaths which make no sense unless we use Psycles, recognize the function of subconscious identifications, and realize that there is also subconscious communication which permeates our lives.

In an industrial "accident," 9 A.M., Monday, August 24, 1959, Mike Sanger severed his jugular vein and bled to death. Saturday evening, about 36–37 hours before, his closest friend John Jacoby was killed in an auto wreck.

In another industrial explosion in a chemical plant manufacturing plastics, a worker "accidentally" touched off a bucket of acetone which had spilled over him, and he raced around as a "flaming torch" to his death. A Psycle earlier, his roommate (lover?) had packed up and left after a heated argument.

A wealthy socialite "with everything" was found dead in a motel room in Miami Beach, Florida, January 23, 1970. Psycles before, she had attended a friend's funeral in San Francisco. Was there some trigger there, to which she responded with her own death?

Where two rivers flow together to become one river, it is a confluence. So it is between people and their needs. There is a natural flowing together, a rightness and cooperation despite all surface appearances of conflict or "misunderstanding."

In Orlando, Florida, on Saturday evening, November 29, 1975, Frank D. Booth, Jr., age 50, pulled to the side of the road and was parked on the shoulder of the highway. Trooper Robert

Rennie, age 25, believing it to be a stolen car, approached, thought he saw Booth reach for a gun, pulled his own, and shot Booth dead.

As it turns out, Mr. Booth was on his way to the airport to fly to Northhampton, Massachusetts, to attend the funeral of his father, Frank D. Booth, Sr., who died Friday morning. As we see it, Trooper Rennie was the instrument of Booth's suicide. In responding to that command signal, the policeman probably fulfilled also a need of his own to kill "father." Hence the confluence. We know what could have triggered Booth's death, but not the trigger motivation for the cooperative policeman.

And so on, and so on. Just underneath the surface, somewhere, and especially with the help of Psycles, we can find the precise reason for every death. There is motivation, and a decision to die in a particular manner.

Karen Ann Quinlan, the famous legal case in which the parents sued to be permitted to "pull the plug," is now in her fifth year in a coma. In a poem she wrote just months before her coma started (April 15, 1975), she left a clear message to explain her action:

"The constant struggle with submission is tiring; the so-called strength I've gained is just another heavy load; I wish to curl myself into a fetal rose and rest in the eternal womb awhile."

As a 70-pound vegetable whose brain is said to be damaged beyond repair, she is still, nevertheless, *doing* what she *intended* to do.

Sudden Infant Death Syndrome (SIDS)

Every year over 10,000 healthy babies suddenly die in their cribs without good reason. Most of them (90 percent) die in their sleep between the first and seventh month; most occur in lower socio-economic families; most of them show some signs of respiratory infection but not of such severity as to cause death; most of them have mothers who smoke; there is no obvious hereditary connection; it is a full-blown mystery with many investigators.

As usual, many causes are proposed, none of which are psychological. Some researchers are looking for viruses, others study the sleep process and the periodic "apneic episodes" in which breathing just stops for a few moments and then usually continues. Some look for pathologies, such as "abnormal neurological messages" between the brain and breathing reflexes, or abnormalities in the development of nerves controlling the larynx, or abnormal muscular spasms, and so on. One doctor recommends that babies sleep at a slight incline to prevent obstructions to breathing that might be caused by puddles of vomit. Another doctor blames botulism from spores from honey.

And then there are the expensive electronic monitoring machines that sound a warning signal if the baby's breathing stops.

The emotional turmoil around these deaths is great. The parents usually have strong feelings of guilt, certain that they are somehow responsible for the baby's death, wondering what they might have done differently. Some parents won't even return to the home where the child died. With others it is followed by divorce.

If you have followed the massive Psycles evidence of self-responsibility in the psychogenesis of the pains and distortions of disease and mis-behaviors, you will anticipate what we must conclude about this problem.

It is quite simple: SIDS babies should be blessed as little martyrs who did the right thing, gracefully stepping aside and bowing out of a stressful situation. Consider these premises for what could be a motivated suicide:

1) Very few babies are planned; most of them are "accidents." If the mother desperately needs income from work, motherhood brings an economic crisis. (Thus: the 90 percent from lower socio-economic families.)
2) Parental stresses, tensions, conflicts are *felt* by babies; they may cry incessantly and the agitated mother may feel hopelessly unable to cope. Most parents can admit that there were times when they felt the wish that their child would "drop dead." The powerful biological instinct of motherhood, however, usually prevails over these death-wishes.
3) There is no question but that babies are deeply aware of

> being wanted or not wanted. With an unmistakable signal of total rejection, the baby may confluence and agree to die by turning off its heart 37 hours (etc.) later.

In the case of the Espiritu child mentioned on page 159, the total rejection involved violence and injuries—but these were not lethal. The child went to bed *two* evenings before ending his own life about 37 hours later.

An alternative scenario could just as easily have been clamping a pillow down over a screaming baby just long enough to convey the message, but not long enough to suffocate. The child is released, revives from near-suffocation, and then exits life at the same 37-hour time-lag without evidence of physical damage. Or, the child may be screaming nonstop while the parents sit in another room, slam the door, feel terribly hostile thoughts as they decide to "just let it cry it out." During that time the child again *feels* the total rejection, and again makes the decision to exit, unwanted, 37 hours later without signs of physical violence.

Is all this so difficult to confront? Some people have argued that piling guilt on guilt would be too horrendous. But I would assert that the parents, deep within, already know it. So it would be much healthier, psychologically, to admit the truth and open their hearts to the underlying reasons why they wished the child dead—even if it was only a temporary feeling. It may be a shock at first, but far healthier in the long run not to play the role of innocent liars. The secret ties that bind, the identifications, the problems, confluences, and agreements are all there, if we look, and the looking would be fundamentally liberating.

Again, I say—*bless* the child as a little martyr for stepping aside to relieve a burden. And then quit the game of self-punishment and remorse.

Life *is* marvelous. Life *is* wondrous. And things happen rightly!

A Better Way of Dying

When someone is dying, most people don't know what to do, how to relate, how to communicate, or how to bring beauty to the whole scene. We wish we could wave a magic wand and bring

Soldier's Wife Charged in Son's Death

By Jack Hopkins
P-I Staff

TACOMA—The 23-year-old wife of a soldier was charged in federal court here yesterday with beating her 2½-year-old son to death last week in the couple's apartment at Fort Lewis.

Deborah Renee Espiritu cried softly as U.S. Magistrate Robert Cooper read the federal charge that she "with malice aforethought and by means of a beating, unlawfully killed Derik K. Espiritu, her child."

The frail young woman's husband and other family members looked on as Cooper ordered her held in lieu of $25,000 bond and set *Feb. 27 for a preliminary hearing*.

The child was found dead at the couple's residence, 6490 Garcia Blvd., Saturday morning by special agent Mike Farley of the Army's Criminal Investigation Division. He was there for what had been *reported* to him *as a crib death*.

The child was taken to Madigan Army Medical Center where he was pronounced dead.

Examination of the child showed he had two skull fractures, two black eyes, bruises on his shoulder, tongue lacerations and a split lip, according to the federal complaint.

Cooper was told that the FBI questioned Mrs. Espiritu that afternoon and she gave a *statement* to them after being advised of her rights.

"Deborah Renee Espiritu stated that on Feb. 14 she was home with her child, Derik. The child was on and off of a rocking horse while she was busy writing and doing things around the house," according to a signed statement by the FBI.

"The defendant stated she started to get mad, hit the child by slapping his face, picked him up and shook him, and in the course of shaking him, hit the child's head on the flat wall of the living room.

"Mrs. Espiritu said the child hit his shoulder on the rocking horse as he fell.

"She sent the child upstairs to his room and the child *seemed normal*" the FBI said.

"The child went to bed, according to Mrs. Espiritu, at approximately 7 p.m. on Feb. 14.

"The child got up the next morning at about 6:30 a.m. Mrs. Espiritu stated that she only hit the child on one occasion on Friday Feb. 15. She stated that the child's *appetite seemed* bad, that he *seemed sick* and that he was *breathing unusually* and she thought the child had *congestion*.

"On Saturday morning, Feb. 16, the child did not wake up and was dead. The child did not receive medical attention between Feb. 14 and Feb. 16," the FBI said.

Mrs. Espiritu remained silent throughout the brief hearing.

Cooper said he would appoint an attorney to represent her after her husband—an enlisted man—said he only made $600 a month, the couple had about $200 in debts and couldn't afford to hire a lawyer.

Following the hearing, she embraced family members and wept before being led away by federal marshals for the trip to Pierce County jail.

THE STRESS (TOTAL REJECTION)	→ Thurs.	aft. (time?) 7 P.M. to bed
	Fri.	6:30 A.M. up mild symptoms 7-8? to bed
Suicide Response C.37 HOURS LATER	→ Sat.	early A.M. 6-7 A.M. found dead

PSYCLES IN THE NEWS: This case is similar to what is known as SIDS (Sudden Infant Death Syndrome) or "Crib Death," except that the trigger incident (the total rejection) was violent and left physical marks. (Seattle Post-Intelligencer, Wed., Feb. 20, 1980) instant healing, comfort, and happiness, but we feel helpless to do so.

Despite our good intentions, and because of our real fear of death, we find ourselves caught in a web of deceit, pretenses, and hypocrisy that extends beyond the family to include the nurses and doctors to whom death can be taken as a personal or professional failure.

The denial of death permeates our whole society and many studies have shown its insidious effects. We shun. We cover up. We avoid and try to skirt around its edges with as little contact as possible. There is avoidance, isolation, and the agony of loneliness. All this goes terribly hard on the dying.

It should not be this way.

Death can and should be an event which is beautiful, friendly, peaceful, painless, and ennobling for all concerned, both the dying and the surviving. Remember our case of the 96-year-old grandmother?

We can't change the past, right all the wrongs nor transform a whole lifetime of actions and inactions. But from where we are, right now, we can bring about changes in attitudes, feelings, and relationships to make the departure much more pleasant. Even beautiful.

Good advice toward these ends can be found in the various books on death and dying such as *On Death and Dying* by Dr. Elizabeth Kübler-Ross, *Learning to Say Good-Bye* by Eda LeShan, *How Could I Not Be Among You?* by Ted Rosenthal, *Man's Search for Meaning* by Viktor Frankl, and many others.

Don't talk down to the dying. After living all their lives with adult responsibility, they deeply resent being treated as children.

Don't be so dishonest. Almost all persons approaching death wish the truth. They want a more honest relationship, a more candid discussion of realities. The feelings of uncertainty, helplessness, and loneliness which come inevitably from deception, false encouragement, and dishonesty may be far more difficult to bear than the truth of impending death.

Don't underestimate their knowledge. People have a way of knowing far more than we give credit. There is a subconscious knowledge that goes far beyond what people tell each other. We know what we're doing to our bodies, cell by cell. And all the pretenses, lies, and avoidance of truth are strictly games.

Don't waste, therefore, the precious time remaining. Such foolish deceit merely deprives the dying of the time they need to attend to the many things they must do to prepare themselves for their departure. There are practical matters such as settling disputes, making wills, mending fences, and reestablishing good will, discussing financial and other property matters, and any other unfinished business. Leaving all this undone can be a massive barrier to a joyous departure.

Don't always insist on being cheerful. There are many feelings and emotions which the dying need time to express. Trying to cheer them up may therefore interfere with their going through certain necessary changes. Dr. Elizabeth Kübler-Ross describes in detail the following five stages which the dying go through, although not necessarily in the order given:

1) shock, denial, and a feeling of isolation
2) anger that this should be happening to them
3) bargaining for more life by promises
4) depression in facing and mourning own death
5) acceptance, and the severing of ties.

We should recognize these various aspects as normal functions of the process, and be neither surprised nor opposed to their expression. The dying need to feel grief over impending loss of treasured relationships, regret over failure to fulfill certain hopes and dreams, anger over their helplessness to change things, and remorse over actions or inactions where there may be feelings of guilt. Give support to their expression.

Don't insist on heroic measures. Dying by inches, in misery, despair, and isolation, held by on borrowed time by artificial means, with tubes in and out of every body opening connected to machines and monitoring instruments—is this really being alive?

The Hospice Program

Throughout the United States this past year there were over one hundred hospice programs in various stages of development, designed to bring loving care back to the dying.

A directory is being published by the national organization of hospices at 765 Prospect Street, New Haven, Connecticut 06511. The New Haven Hospice, first of the American programs and the most advanced, was guided by Dr. Kübler-Ross. We recommend you get in touch with them for help.

Although modeled on an original hospice program started in London, there are a number of different versions growing in the United States. In some cases the program centers on a special wing of a hospital dedicated to humane dying, where family members are encouraged to join with the medical staff in caring for dying relatives, with unlimited visiting hours, and where children and pets are free to visit.

In other models, teams of nurse-therapists are on 24-hour call to make arrangements, where possible, for the dying to die in their own homes rather than in the impersonal atmosphere of strange hospitals.

Of paramount importance is the alleviation of pain so that the patient can attend, without distraction, to the many matters needing attention prior to departure. Liberal use is made of a "Brompton's Mixture" consisting of morphine, cocaine, alcohol, syrup, and chloroform water. This dulls the pain, making the patient much more comfortable, while inducing a little euphoria, which is nice, without dimming alertness.

You will remember in Chapter 3 we described our self-image identifications as "conditional," and in Chapter 4 our psychosomatic ills as "conditional." Is there an "un-conditional" state that we may touch on in this area of dying?

The Catonsville Research

In the March 1973 issue of *Harper's Magazine,* beginning on page 56, you will find a deeply moving article by Jerry Avorn called "Beyond Dying." It describes years of research by the Maryland Psychiatric Research Center in Catonsville, Maryland. Under grants from the National Institute of Mental Health, terminal patients were administered various psychoactive chemicals designed to ease the agony of terminal illness.

In the words of the psychiatrists who have guided almost one hundred such sessions, some strange things happened: "Not all patients reach transcendence. A majority experience some kind of heightened insight into the meaning of their death even if it falls short of cosmic consciousness . . . for those who undergo the destruction-rebirth-cosmic unity experience, death, instead of being seen as the ultimate end of everything and a step into nothingness, appears suddenly as a transition into a different type of existence.

"The patients who have transcendental experiences develop a rather deep belief in the ultimate cosmic unity of all creation and experience themselves as part of it without regard to the situation they are facing."

These psychiatrists are describing a deeply spiritual experience, which we believe to represent the temporary induction of the unconditional state. It was as if these dying patients shed for just a

A medieval artist illustrates the breakthrough from the conditional to the unconditional.

moment their lifelong conditional self-image, and found themselves suddenly free in a wondrous cosmos. There is no question but that it was immensely beneficial in dispelling some of their fears of dying.

Can their experiences help guide us toward new insights about death?

Some Speculations on Death

Let's consider some of these clues together, and see where they might lead us.

1) Psycles shows we are conditional (our self-image, identities, and sufferings).
2) Psycles also shows that freedom and health can come from deconditioning.
3) All therapy systems can be considered as methods of deconditioning.
4) The Catonsville research shows that people can experience the unconditional, and with a little help from chemistry, bring about a cosmic consciousness event.
5) A classic book by Dr. R. M. Bucke, a Canadian psychiatrist, *Cosmic Consciousness*, describes many unconditional experiences.

Now, if there is the conditional and the unconditional,

And if, as Psycles shows, death is intentional and self-responsible,

Then, we can conclude:

1) We can't decondition death itself; we are all mortal, and must die sometime.
2) But we can decondition certain specific identifications and self-images which portend an early death through accidents or diseases.
3) Deaths do not occur as the result of one "trigger incident," but evolve out of a whole identity structure and self-image and its experiences.
4) With everyone there are plenty of warning clues that appear from time to time of events which transpire, leading up to the actual decision when and how to die.

5) There is a "rightness" and "appropriateness" about all deaths. No one dies against his will. (Of course, we mean generally subconscious.) Though we may rightly mourn their loss, there is nothing "wrong," bad, wasteful, etc., about death.
6) Though we may think it *could* have been different, the image *could* have been changed, and they didn't *have* to die thusly, we must recognize that they chose the how and when of it, for their own reasons, and chose not to change that choice. We must accept the "rightness" of all that.
7) *At death:* All dying persons *experience* "cosmic consciousness," a merging with the infinite, a "return to the source," the "transcendental," a "union with God," the "destruction-rebirth-cosmic unity," that many living persons have experienced. Basically, they surrender the conditional identity-game, the self-image, and transcend to become free and unconditional.
Realize: that the cells of the body are still very much alive long after the heart, breathing, and brain-waves have ceased. I would conclude that throughout rigor mortis (which lasts about 24 hours), that "person" is in a state of ecstasy, fully "alive" in trillions of cells. When rigor mortis stops, and the body relaxes, the molecular structures are then allowed to fall apart, disintegrate, decompose. The "person" is now "one with God." Only "God" remains, omnipresent.

In this chapter we have shown how Psycles can bring new illumination to the whole subject of death. Instead of the illusion of people being seized by an evil force and carried off against their will, we find the miracle of people choosing the time and manner of their departure. We see something of the motives involved, and in Chapter 16 we will discuss how the defusing of this fearful image can lead automatically to a new and healthier life.

Now let's see how Psycles can dispel one of the most ancient of myths—the innocent victim myth—and bring new meaning and hope to our judicial system with its problems of dealing with crime and punishment.

10 Crime and Punishment

One of our deepest illusions is the idea of the innocent victim. When a person commits a crime against another person, we wish to believe that the first person is "guilty" and the second person "innocent." We are taught it from infancy; everybody believes it. So to suggest that a victim *wanted* whatever it was that happened to him or her may trigger outrage in some people.

"Nobody's gonna tell me that I *wanted* that!"

Even when we say that it's the subconscious that we're talking about, they might snarl, "You think I'm crazy or somethin'?"

No. This is all very "normal." It's when we use Psycles time-lags to analyze the prior causative events for *both* the offender and offended that we almost invariably find that both are equally responsible! There is a flowing together, a confluence of needs.

When a woman is raped, we can find out why it is that she wanted and needed to be raped (subconsciously) in the events of her life Psycles before. Again, we run into the inescapable fact of subconscious communication. The offender and offended come together by mutual consent.

As Kahlil Gibran wrote in *The Prophet*—

> "And this also, though the word lie heavy upon your hearts:

The murdered is not unaccountable for his own murder,
And the robbed is not blameless in being robbed.
The righteous is not innocent of the deeds of the wicked,
And the white-handed is not clean in the doings of the felon.
Yea, the guilty is oftentimes the victim of the injured,
And still more often the condemned is the burden bearer for the guiltless and unblamed.
You cannot separate the just from the unjust and the good from the wicked;

. . . .

If any of you would bring to judgment the unfaithful wife,
Let him also weigh the heart of the husband in the scales, and measure his soul with measurements.
And let him who would lash the offender look into the heart of the offended.*

Because we don't wish to accept what has been stated so gently and poetically by mystics, we have ignored this fundamental truth. What would happen to all the people we *blame* for things? Where would we direct our hate?

Some years ago a staid college professor, Dr. Dement, performed an act which was otherwise inexplicable and caused him a great deal of embarrassment. While walking along a city sidewalk he had a sudden irresistible urge and proceeded to use his foot to smash in a number of store windows.

This was a crime. He was arrested. He had to pay the penalties. We can assume from our research that what he did was homeostatic. This explosive act presumably "balanced the books," quite possibly for a severe conflict or break in personal relations with some wife/mother/female. The building was a big mother he was doing in.

But consider *also* the needs of the store owner or manager or display artist. We would expect to find a need there, also, that had to be fulfilled: something dramatizing paranoia, perhaps, being attacked, exposed, broken into.

*Reprinted from *The Prophet*, by Kahlil Gibran, by permission of Alfred A. Knopf, Inc. Copyright 1923 by Kahlil Gibran and renewed 1951 by Administrators C.T.A. of Kahlil Gibran Estate and Mary G. Gibran.

We would predict that insight into the professor's act (by making connection with its triggering cause) would mean that he would never be impelled to repeat that act again. It would open up one pocket of his occluded subconscious; he would be healthier, less tense, more aware of certain solvable problems, and less liable to do something similar in the future.

Can you visualize the entire world, with all of its trillions of interactions, all happening in an integrated fashion—all mutual, holistic, and magnetic in its overall oneness?

A thief who steals, an alcoholic who starts a barroom brawl, a motorist who smashes his and someone else's car, a flier who crashes his plane, an arsonist who sets a building on fire, a pervert who molests a school girl . . . all of these can be unlocked by the Psycles Master-key to show their needs, motivations, and the cooperation of all involved.

We have already shown that death is no exception, and how murders must represent suicides on the part of the victims, in a cooperative coming-together involving subconscious communication. Remember the theater usherette who was run down on the sidewalk in accordance with her identification with an older brother? That stranger who "accidentally" let his car go "out of control" *had* to be in communication with her to do what he did.

Thieves don't steal from just *anyone*! Their "victims" are only those who wish to be robbed.

In soliciting TV time for Psycles, I sent a flyer to a San Jose station talk show and got the following phoned response:

"Your flyer listed robberies among your 2,000 case histories. It so happened that my secretary came in this morning saying she had been robbed. So I called her into my office and asked her when she discovered it. We then figured back 37 hours (with some trouble, I might add), and when we landed on the time, and I asked her what had happened at that time, she said, 'I don't want to talk about it!' So, Mr. Bulkley, it works! Come on down and let's talk about Psycles on my show."

Robbed by a Guru

In June of 1974 while visiting in India, I came upon a marvelous example of the perfect cooperation between people in crime. One adventurous psychologist, after weeks of hesitation and much soul-searching, attempted to perform a supreme act of propitiation to his guru. He was rejected and felt it so deeply he contemplated suicide. Thirty-seven hours later (almost to within minutes!) he was severely shaken by the discovery that he had been robbed. A thief had lowered a long hook through a hole under the eaves of the adobe hut where he lived, and pulled out a bag that contained the devotee's airline tickets, his glasses, and wallet.

The amazing thing about this theft is that those particular items featured prominently in the decision-making problem: should he continue writing a certain book or should he return to England where he had a speaking engagement? (The items, incidentally, were recovered by the police.) How did the robber get just those particular things? The psychologist had put them selectively in a small pouch and placed them against the wall (all subconsciously) in a manner perfectly suited to the theft.

And the motives? Here we have the 37-hour timing between the psychological feeling of "being robbed" (denied what he felt were his just deserts from the guru) and the overt manifestation. The crime was done in accordance with the Englishman's subconscious command signal. He needed to be robbed. A robber cooperated, taking only the correct things which highlighted his what-to-do problem.

This again implies a species of omniscience, of communication linking bodies intimately—everywhere.

The Little Old Lady—Mugged

The classical case of the "innocent victim" is the elderly lady who is walking along, minding her own business, when some very bad person suddenly deprives her of her purse. If the victim should be poor, clutching to a Social Security pittance, it becomes all the more difficult to see that she may have *wanted* to be robbed.

In our mugging case her husband had warned her, "It's getting dark. Better that you should wait until tomorrow." But on the evening of Friday, March 17, she had gone anyway, walking several blocks to the local market to do some needed shopping.

Returning about 8 P.M., well after dark, she was suddenly

confronted by some "punk" who grabbed her purse. After a brief tug-of-war which spun her around and down to the sidewalk, he was off into the night chased only by her screams. A paranoia quickly enveloped her household as police were called, reports filed, credit cards notified, and all the locks on the doors changed (the purse contained their keys). Three days later the purse found and returned, minus only the visible money. Much to her glee, a secret compartment containing a larger sum of money was untouched.

Now, we ask, what would motivate such an event? What would make her *want* to be mugged? The key to it all was her birthday, Wednesday, March 15. It had always been an important affair to her, But Tuesday evening (the three-day period before the robbery) she was deeply disappointed because plans were canceled; her husband was sick with stomach flu and her daughters had busy appointments. All Wednesday she felt "robbed" of her customary party and the special family attention she longed for. Friday evening that suppressed feeling was out in the open due to the cooperation of a mugger who subconsciously tuned into her signal.

The Secretary Who Was Attacked

In a downtown office building, about 11 on a Tuesday morning, an advertising executive sent his secretary down to a lower floor to deliver a proposal to a client. Having presented the papers, the secretary suddenly found herself being chased around the desk by a man in his early sixties who had been the very model of propriety but was now energetically trying to hug and kiss her.

She broke loose, went back upstairs to her office, and told her boss what had happened but added that she didn't want to press charges or make any trouble, because the client was really a decent senior citizen gentleman.

The boss, familiar with Psycles, thought: "Ha! Let's see whether Bulkley's theory holds in this case," and asked, "What were you doing Sunday evening about 10?" (That would be about 37 hours earlier).

A few puzzled frowns later, she was unfolding the following story: Sunday evening she was feeling especially loving toward her husband ("horny" was the word she used). He dropped into bed early, while she took a shower. After the shower she took extra care fixing herself up in a sexy black negligee, added touches of his favorite perfume, came out, and found him sound asleep. She touched him gently. No response. She snuggled up to him in bed. He was still dead to the world. Unrousable. She lay there angry and frustrated. And this was the repressed signal which erupted 37 hours later to trigger the unfatherly advances by the client.

We can only assume that he had his own frustrations—whether Sunday, or another Psycle back.

Several Cases of Rape

In the only case of rape thus far analyzed by the Psycles method, a dancing instructor was attacked and raped in a dark alley in a part of the city of Cincinnati where she had never previously been. When the husband and wife were questioned as to what happened 37 hours before the incident, they looked at each other, their faces went white, and they said they both "saw" it (the logic of why it happened) but refused to discuss details.

In Buena Park, California, newspapers reported two separate incidents that occurred on January 11, 1960. The fact that one teenage girl was raped at 4:40 A.M., and another teenage girl was abducted at 5:15 P.M.—exactly twelve and a half hours later—suggests a common trigger, such as a late-late TV movie, for instance, Thursday at about 2 A.M.

The Need to Be Punished

Officially, the *causes* of crime are often declared to be due to unemployment, discrimination in schools, the mobility of populations, the increase in high-rise buildings, the increase in juvenile population, the economic situation, lack of opportunities, and even to examples of governmental and political corruption. We are generally urged to make the criminal justice system quicker and more efficient, to improve our incarceration and rehabilitation programs, to put the "truly dangerous" away, and increase penalties for repeaters.

But where does all this stand when we look at each case

individually and find confluence, communication, cooperation, and the fulfillment of needs everywhere? Some very perceptive articles have been written about the interlocking bind which exists between addicts and narcotic agents. The game-structure and the cooperation and mutual dependency show up clearly.

If it is true that crimes between persons are cooperative events when viewed from the standpoint of subconscious realities, then what does this say about the whole process of justice? Is it truly blind?

On the one hand you might say that the entire judicial system is a farce because a) we all know, subconsciously, who did whatever was done and why, and b) neither party was exclusively "guilty" against the other party's "innocence."

But going one step further, it can be better said that the criminal offender was transacting *for* being caught and *wanted* to be processed and *punished.* You will recall some of our "accidental" case histories and how they show the fact of self-punishment. In criminal cases, part of the need for the whole drama of arrest, trial, and punishment may have deep symbolic meaning in terms of the equivalence between father/God and law-and-order. When the law finally catches up with you and incarcerates, it is like being swept up into daddy's arms. It may be painful, but it is also secure and full of attention.

Something like this happened to me, years ago, when my company merged with another company and a hatchetman came in and wiped out the top managerial staff, giving me one day's notice. I felt very insecure. Thirty-seven hours later, late at night, I went through a four-way stop, without fully stopping. It was dark, and there was no traffic. But there was one lone policeman waiting there for me, out of sight. He gave me a ticket.

The timing meant that I wanted that sort of attention, that I knew where a cop was waiting, and that I deliberately (though subconsciously) went through that intersection in order to get what I got. Maybe you can think of a better interpretation, but it *seemed* to mean that I wanted to feel the arms of the law, like daddy enfolding me, way down deep. Actually, in much of our research and case histories, it would seem that the interpretation is not too important in every case. Making the connection is. And the conscious putting of the two and two together, the connecting, automatically sets wheels in motion in some manner which is liberating.

Police Brutality Is Also Cooperative

In two cases of police brutality analyzed with Psycles, the motivations on the part of both "innocent victims" stand out clearly to show that both got what they wanted, and that the police were being duly cooperative with their needs.

In the first case the police came to break up a parking lot brawl in Laguna Beach, California. They intended to take our subject, only 15 years old, home. But by provocative language he induced them to necklock and handcuff him and throw him in jail. Why did this teenager deliberately provoke the police to manhandle him? Precisely 37 hours earlier, working for his tough, ex-marine father, he had been severely criticized for some shortcoming which he felt was totally unjustified. His impulse to sass his father, repressed, was let loose on the cops, who then "enfolded" him with a physical contact he never got from his father.

The second case was a 25-year-old man who was separated from his wife but hanging on, pestering her with much unwanted attention. About 5–6 P.M. on Sunday, there was a fast dialogue in which he was accused (quite unjustly, he felt) of preventing her from completing her Ph.D. thesis which was due that evening. Thirty-seven hours later, Tuesday morning very early, after being arrested on his wife's complaint, he provoked a violent scene ("I'm no criminal!") and was processed and booked for a ten-day stint in city jail.

Psycles and the Future Judiciary

With support from many independent observations, and finding validity in its therapeutic benefits for a healthier and freer life, we are nevertheless not so naive as to assume that this data will have much immediate impact on our way of life.

For the most part, worldwide, we prefer to think of ourselves as innocent victims of a host of external forces and circumstances, rather than as responsible agents for whatever "befalls" us. In spite of clues from psychologists or mystics, we prefer to blame things "out there" for our troubles. The truths at the roots are essentially painful.

Looking into the future, it is easy to see that certain changes are in order.

First, of all, with the various instruments current and soon to be available (e.g., the lie detector known as the Voice Stress Analyzer), it will be possible to have a fairly clear picture of the truth of what actually happened, including motives involved, without going through the blindman's buff game in which all pretend that nobody knows anything.

Dispensing with the fiction of the aggressor/victim adversary relationship, in which one is to be punished and the other rewarded with compensation, we establish that both parties wanted what happened, and that it was a cooperative 50/50 dance between conditional sufferers with mutually interlocking aberrations.

We then proceed to decide what to do about it, with these possibilities:

1) both parties should redress damages to the best of their abilities: the "aggressor" who transacts for prison should of course get what he was asking for;
2) but neither party should profit financially;
3) both parties should become involved in seeking a solution for their aberrant motives;
4) if the injured party is incapacitated for useful employment, it should be seen as a willed self-destruct; (s)he should not receive a multi-million-dollar settlement with all its vagaries of years of care, financial investments, and trustee arrangements, but might instead become a ward of the state.

Generally speaking, disentangling ourselves from the concept of the innocent victim can lead toward a greater sense of social involvement and responsibility. There are invisible bonds which link us together deeply.

11 Dreams and Nightmares Reveal Their Secrets

From the most ancient times, dreams have been a worldwide fascination. During every 24-hour period, bodies get tired, lie down, and then go unconscious in some manner. During that strange interlude called sleep, many internal adventures can happen, some quite terrifying.

Sometimes the dreams are prophetic, as in the classic cases of Joseph in the Bible, and Lincoln's dream of his assassination. A study by a London newspaper turned up many nightmares which foresaw the disaster in Wales, when a mountain of coal slag engulfed a school, killing about 122 children. Many Jews in Nazi Germany had explicit dreams in advance of the Holocaust. The evidence certainly suggests we have some insight into future events.

Dreams can be embarrassing in their conflict with normal standards of conduct; St. Augustine expressed his thanks to God for not making him responsible for his dreams! While there can be frightening nightmares involving fears of being attacked, of dying, falling, being chased or entrapped, many creative persons rely on dreams for insight and inspiration.

Some Recent Discoveries about Dreams

During the last two decades there have been penetrating discoveries about sleep. With the use of brain-monitoring instruments such as the EEG (electroencephalograph), it has been shown that people generally go through a rhythmic cycle of about 90 minutes, which includes several distinct types of brain-wave states. It begins with a Stage I (threshold sleep), goes to Stage II (with sharp spindle waves), to Stage III (with large slow waves), and finally to Stage IV (the large synchronous delta waves). Stage IV is the deepest slumber, the sleep of the weary, which predominates during the early part of the night.

Each 90-minute cycle, the sleeper comes up from delta sleep to what is now called REM sleep in which there is rapid eye movement associated with dreaming. During the earlier parts of the night, the REM sleep stages are shorter, maybe about ten minutes. During the early morning hours, the REM stage takes more and more of the cycle, and one may be dreaming for such extended periods as 20 minutes or longer.

During the REM stage, if the sleeper is awakened, it will invariably be in the middle of a vivid dream. The REM state and its dreaming is apparently necessary. If a sleeper is interrupted each time the REM shows, and dreams are thus prevented, the subject will suffer nervousness, become irritable, lose coordination, and manifest other personality changes.

What, then, does all this tell us about dreams?

A Look at Some of the Theories

For centuries many have thought that dreams were visitations from the gods with prophetic portents for the future; dreams were also considered as the meanderings of the soul released from the body by sleep. At the other end of the spectrum are the biochemists who would prefer to think of dreams as merely products of chemical activities in the brain.

Noting that adults dream less than children, and very premature infants are in a continuous state of REM sleep, some researchers have concluded that dreams are part of the learning process. A person is sorting new material picked up during the day and seeing how it relates to the self. Dreams may represent tiring work, because if you sleep too much and dream too much you may wake up tired.

One theory holds that if people were in the deepest phase of sleep (being deaf, dumb, and blind to the world) for too extended a period of time, they would be in greater danger of being an easy prey, so dreams were evolved as a survival mechanism to keep the sensory apparatus functioning so that they would not be "caught napping."

The symbolic content of dreams has engaged many psychologists as well as psychic counselors. Freud thought they were generally "wish-fulfillment" and saw many sexual repressions and bodily preoccupations disguised in dreams. They would be overt expressions of what were otherwise unconscious emotional conflicts and impulses. Jung said that dreams reveal the unconscious mind, can anticipate future events, and are often shaped by innate structures called "archetypes" originating from a collective unconscious. Adler gave focus to another function, that of compensation.

While most psychiatrists do not agree with Freud that all dreams are wish-fulfillment, many would agree that dreams at least help us cope with problems in some manner not clearly defined. Fritz Perls saw dreams as a method of reclaiming lost parts of the personality to become more whole. While there are many lists of dream symbols and what they may mean, it is more probable that there is no universal symbolism; symbols have individual meaning for people, depending on their experiences and associations. For instance, Freud held that dreams of flying symbolized sexual intercourse, whereas Jung believed that they pointed to potentials for personality changes. Whatever the underlying meanings, dreams represent a great deal of camouflage, and efforts to interpret vary widely.

Before getting into examples of how dreams may be analyzed or programmed for life-enhancement, let's first look at how Psycles can help solve specific problems such as sleeplessness or nightmares.

Psycles and Sleep Problems

For those who have sleep problems, Psycles makes a unique contribution beyond the usual advice by sleep therapists. Among their traditional what-to-do's are the avoidance of excitement (such as reading mystery novels) and excessive sugars just prior to sleep time, listening to soothing music, and the imagining of pleasant scenes associated with feeling secure and tranquil.

By contrast, the use of Psycles adds a whole new dimension to the problem-solving. You remember the case in Chapter 2 in which the engineer was waking tensely each morning about 5, and how this was a reflection of his steering-gear problem driving home each evening about 4? We are no longer dealing in generalities. Pinpointing the specific stress-trigger at a circadian time-lag earlier can make all the difference in the world—the difference between knowing and not knowing why.

The Uptown Nightmare

A Seattle businessman awakened one night with the most terrifying nightmare he had ever had. Struggling up from his bed in a cold sweat he turned on the light and found that it was 1 A.M.. Slowly regaining his composure, he remembered Psycles, and checked back 37 hours to find the exact reason why.

At noon, a day and a half before, he had been in the northern part of the city with a client when he picked up the phone and called his office. A catastrophe had occurred. Whatever happened (he didn't specify) had almost destroyed his business. And he spent the best part of the next week trying to pull the pieces back together again.

Now, he might have gone back another 37 hours to get some insights into why the catastrophe happened, but his failure to do so probably reflects the likelihood that that would have been more than he could handle at the time.

Three Sleepless Nights

At a Friday evening social, a young speech therapist named Betty was saying that Thursday night was the first decent night of sleep all week. "Monday night I worked late, puttering around the house until 3 A.M. restlessly. I went to bed, but couldn't sleep, tossed and turned until dawn. Tuesday night, still no sleep. And Wednesday was again a restless night."

"So it all started Monday night?"

"That's right."

"Anything bad happen Sunday afternoon?"

"Sunday wasn't so bad. But Saturday! That was terrible. I went to a picnic at our ski club but nobody was there. So I decided to phone friends to find someone to do something with. Made twelve phone calls to friends in the Seattle area. No one was home. Then I tried the Tacoma area. Still couldn't contact anyone. I finally curled up in bed and read a book, feeling terribly alone, isolated, and rejected."

Why the sleepless response started Monday instead of Sunday night remains part of the Psycles unknowns. You will remember in our illustration in Chapter 1 that the time-lags are multiples of 12-plus hours, with a preponderance of cases at 37 and 74 hours. This two and a half day lag is more rare, but draws attention to the need to check other nodes if you don't hit "pay dirt" the first try.

In a further commentary on this case, Betty recalled, "As a young girl I once told my mother about a sleepness night and she groaned 'Don't tell me there's going to be *two* with sleep problems in this family!' You see, my father was a very light sleeper, and would wake instantly, for anything. This annoyed my mother. I guess I was taking after my father."

The Bandicoot Nightmare

During my late teens and twenties I had a nightmare which roused me in a cold sweat every now and then, but I could never lay my hands on it. The content remained a nameless terror that always skittered away into nothingness. Finally, I remember making the decision that I wanted to remember whatever it was that was disturbing me. Some months later I literally grabbed it by the tail and terminated the whole problem.

Its origins lay in a deeply horrifying incident that happened to me when I was going to Kodaikanal school high in the mountains of south India. I was about nine years old, when, one day a shout went up that they had trapped a bandicoot in a small empty back storeroom. (A bandicoot is cousin to a large possum.) About six of us closed ourselves in that small room and, as the terrified animal raced back and forth between our legs, clubbed it to death. After a while I walled that memory away in my subconscious. But it was not forgotten. Periodically, it would erupt in these terrifying nightmares. But once contacted, and brought back to consciousness, it was no longer a problem.

This illustrates something of the basic mechanics of therapeutic "insight." Various "diseases" and troubles discussed in Chapter 3 can be thought as equivalent to a nightmare. They, also, can be solved with insight into their origins.

Using Dreams for Life Enhancement

Beyond the need to solve actual problems having to do with sleep, there are other reasons for attention to dreams. Some people pursue dreams for the pure enjoyment of it. Even without serious effort to analyze their meanings, there can be drama and excitement in the pastime of sharing dreams.

On the more serious side, however, remembering dreams, and even programming them, can be highly beneficial. It can provide unique insights for problem-solving as well as promote greater self-awareness, creativity, and even business success. In both the interpretation and the programming, Psycles makes a special contribution in advance of current knowledge.

Here are our recommendations:

1) Keep a dream notebook or tape recorder, with watch and flashlight, at your bedside.
2) Before going to bed, do some light relaxing exercises.
3) In bed, repeat to yourself, "I wish to remember and record my dreams whenever they happen."
4) When you awaken with a dream—don't move. Lie still, in the same position, with your eyes shut, and mentally review the dream. Then record it, including the time it happened.
5) To program dreams: Sit relaxed, with eyes closed, at about 12+ or 37+ hours before regular waking time. Say to yourself, "I wish to dream an answer to the following problem. I will remember and record it." Focus on the problem, exactly what you want insight on, and repeat that declaration. It may have to do with business, social, psychological, or even philosophical questions. Anything goes!
6) Be patient, as it may take a little time and practice.

From Identity Conflict to Dream to Depression

Here's a letter I received:

"Dear Mr. Bulkley: I am very excited with the possibility that your 37-hour cycle is the key to understanding dreams. Here is a chain of events from my own experience—which was like finding all the pieces to a crossword puzzle:

"I was suddenly depressed on Saturday afternoon, February 26th, at about 2 P.M.. My depression appeared to be triggered by a conversation dealing with aspects of my background—aspects of behavior traits in my parents that I didn't tell anyone else about but which I feared I could inherit or be like.

"Thirty-seven hours earlier, at about 1 A.M. Friday morning, I awakened with a dream in which I had buried my parents in my backyard. No one knew it. The dirt in the garden area was wearing away. There was a round flower bed and I could see the roots below becoming exposed. I knew it would be just a short matter of time before the coffins would be discovered. In my mind's eye, I could see them as being silver.

"Thirty-seven hours before the dream, I was sitting in a beauty parlor having the dark roots of my hair touched up, as I recently changed my hair from its natural dark brown color to blond.

"From this chain of events, I can see that my decision to go blond was a subconscious attempt to be different from my parents (who both had dark brown hair) and when I had it touched up, I was trying to cover up my roots (heritage, or who I really am). Knowing the idea of dying my hair would be upsetting to my parents, brings to mind the expression "They would turn over in their graves." (Smile).

"In short, it gave me a lot of insight into myself." (Signed) Paula M. Craig.

The Hot Foot Dream

It was during a Tuesday evening party in Seattle that a deeply mystical-looking young "psychic" was telling of a dramatic dream in which she was forced to walk on red-hot coals. A sinister Fu-Manchu oriental, accusing her of being a CIA agent or working with the drug-dealing Mafia, seized her by the shoulder and thrust her onto the hot coals. As her feet burned she screamed in anger at the false accusations.

Various persons offered interpretations: "Maybe you're about to be offered an important assignment." "Maybe your psychic knowledge has offended someone on another plane." "You're about to be tested."

"When was the dream?" I asked.

"Yesterday morning, about 5:30," she replied.

"Do you remember what you were doing Saturday afternoon about 4?"

"Oh! That was when we got a flat tire; tried removing the wheel but my lug wrench wouldn't turn the nuts, they were frozen. We called the AAA tow truck. They couldn't get the wheel off either. We were towed to a station and *their* compressed-air power wrench couldn't do it either. Nearby was a Firestone agency. They couldn't get the wheels off. Apparently all four wheels had been cross-threaded when we recently purchased four new tires at Goodyear, and all were in danger of breaking loose. We were furious. Finally we got Goodyear to come and tow the car away. They fixed the condition but denied responsibility, said they had no trouble with the wheels, presented us with a bill which we refused to pay. They finally settled for a small modest fee."

Is there a symbolic connection between hot feet and wheels?

Playing with dreams can be great fun, and with a little imagination there can be exciting interpretations. But sometimes, with the use of Psycles, the real reasons may turn out to be more humdrum than dramatic. But even in the humdrum there are the opportunities for new insights which can be liberating. In this case her elderly mother was the key to the whole event. Both her hips were disintegrating. She had one artificial hipsocket already. More surgery was being scheduled for the other hip. She threatened to become an invalid.

As I see it, the daughter was expressing deep fears in connection with her mother-image identity. Mother was threatening her mobility by the possibility of being physically dependent. Car wheels are like legs, and the problem lay with the connections to the body in both cases.

As it turns out, the grandmother also had troubled hips which were used to control the daughter. Going back to Goodyear, does it not suggest that when they put on those four wheels, they set up the whole scene, subconsciously, by purposefully cross-threading the nuts? Does it not suggest a deep subconscious awareness of the mother's hip problem and her need to manifest it in the car?

How far can we go in this connected universe?

The Seattle Trigger and the New York Dream

There was a time when I didn't know whether time-zone changes would affect response times. For instance, if I have a trigger of some sort in Seattle and then fly to New York or London, would the response (dream or otherwise) still be 37 hours (etc.) later? Or would it be in accordance with the new geophysical coordinates?

It turns out to be the former. In other words, once the trigger is pressed, the response still took 37 hours, no matter where I had gone in the meantime. Here was the actual case:

Sunday noon in Seattle, I visited a home in which the owner had been slowly remodeling, working only occasional weekends. The scaffolding had been up a long time; the place looked trashy. The neighbors claimed it was an eyesore and were threatening a lawsuit. That was one trigger.

Twelve hours later, Sunday midnight, I was watching a TV documentary on the drought, starvation, and death in the Sahara. (This was in April 1974.) It showed one mother, with infant in arms, and breasts that were flat and dry. That was the second trigger.

Monday I flew to New York, where, Tuesday morning, at exactly 4:30, I awakened with a dream which combined both triggers (which were 12 hours apart).There was a litter of pups confronting their mother whose tits were dry. And into the house burst a storm of neighbors bent on cleaning it up and I was shouting, "Get out! Get out!"

Thus the two triggers (12 hours apart) were melded into the one response 37 hours after the first trigger, Seattle time. As a footnote on significance, I might mention that when my father told my mother it was time to wean me, she bared her fangs, and stood her ground with, "You dictated the first four children, but this time

I'm going to enjoy this child!" She continued to breast-feed me for several months longer.

Nocturnal Emissions and Sloppy Tacos

All dreams are not delayed responses to events Psycles earlier. There are well documented cases among the various university sleep-research centers which show that an external stimulus during REM sleep may be incorporated into the dreaming. And somewhere in the classic literature is the case of the woman who dreamed of financial problems relating to "frozen assets" and woke to find that her bare behind was facing a cold breeze from an open window.

In the case of the nocturnal emission, however, it was the 37-hour delay. Our subject dreamed of a Frenchman who delayed his build-up in sexual performance for six hours and then ejaculated for six hours. Waking at 3:15 A.M. with a "wet dream," he traced its trigger to a noontime date with an exciting female 37 hours earlier. They had taken a picnic lunch through the Garden of the Gods in Colorado Springs. In this national monument there is preserved a unique formation of bright red rocks thrusting in phallic splendor out of the ground.

The double-six in that dream may be of interest, because the six-hour timing would be another harmonic of the circadian rhythm, though as we have indicated, the commonest Psycles times for stress-reactions are (in this order) 37 hours, 75 hours and 12½ hours, with scattered cases among the other multiples of 12½ hours.

The "sloppy tacos dream" may be an example incorporating this other harmonic. It took place exactly 43 hours after its stimulus (i.e., 37 hours plus six hours). The eating of the sloppy tacos occurred in Grants Pass, Oregon between 8 and 9 P.M. on a Thursday. During the Saturday afternoon nap, at exactly 3:45 (43 hours later) came a replay with sticky fingers and the same sloppy tacos.

I mention this plebian dream because the whole subject of Psycles is open to much more discovery, and the six-hour harmonic may have to be taken into consideration as playing a role.

Two Predictive Dreams

One subject, actually the wise old-nurse-therapist who gave us the case histories from the Phillips Medical Clinic in St. Louis, sent us one of her dreams because it was uniquely predictive. On November 4, 1952, at 11:40 P.M., she had a dream in which a certain item arrived by mail at 9:40 A.M. Thursday (37 hours later). At exactly that time, a letter arrived informing her that the item had been mailed. (The item itself didn't actually arrive until 9:04 A.M. the following day.) This was sent to us just as a curiosity.

Another person "saw" in a dream, at about 4 A.M., a parking lot accident in which her car was sideswiped, exactly as happened 12 hours later. My guess is that 24 hours before the dream, and 37 hours before the "accident," she had already established that she would have her car hit, exactly where, when, and how, and that the dream was an eruption of her *intent,* one 12-hour Psycle before it actually took place.

Deeply involved in the whole subject of dreams is the larger issue of ESP. In the next chapter we deal directly with the vast unconscious awareness and communication which is shown in many of the Psycles case histories to permeate all life.

12 New Evidence of Unconscious Awareness and Communication (ESP)

IN several earlier chapters there have been hints of the existence of ESP, or extrasensory perception. What we have said is that again and again in our tracking of Psycles cases we have been given precise and positive proof that a vast unconscious awareness exists between human beings, both in contact and at a distance.

In this chapter we will confront the issue head-on. First, we will outline the general conclusions to which we have been led by our case histories. Then, we will suggest some of the reasons why ESP is generally rejected. And finally, we will present some of the typical cases which forced us to these conclusions.

What Is the Nature of ESP?

In our view, no one is an island. We are all connected in an invisible grid which binds together all life. Once we begin to see into the unconscious substratum of existence, we find that events are actually unified in a truly harmonious manner.

ESP, then, refers to a natural, continuous, connective function which is universal. It is not a special faculty or sensitivity which has to be nurtured or developed. That it is mostly unconscious does not mean that it does not govern our lives continuously. All life is totally immersed in this contact (whatever it is, physically) and continuously acting on the basis of this total subconscious awareness, as a species of "omniscience"which permeates our entire existence. In the same sense that Jung talked abut the "collective unconscious," so we say that "everyone knows everything" at the deepest level. Life is unitive, global, divine, and omniscient, despite appearances to the contrary.

There are many people who may accept ESP as a "sometime thing," a little here and a little there, manifesting now and then, like a capability that we may be evolving into, or a capability that we once had, but lost, and can redevelop. But we say that, from our evidence, ESP is not a special ability, nor a special sensitivity; it is not a sometimes on and sometimes off ability. It is an always here-and-now function of all people. It is not more developed in some and less developed in others. It is equal in all persons, and all persons are equally acting upon ESP information and knowledge. The differences between people's apparent ESP is only the extent to which it is expressed. There are various "permission systems" whereby small portions of this omniscience are allowed to manifest consciously. These rules govern its display.

And in the scientific establishment, despite the furtive support here and there, there is a hard core of vigorous denial of its reality.

Why All the Denial?

The answer is simple: the roots are painful. Whether the denial is complete (as among hardcore scientists) or partial (as among those who see it as a "sometime thing"), the same reasons we listed in Chapter 3 for being innocent and non-responsible apply equally to the issue of ESP.

As long as we can agree to pretend that we don't really know what's happening, and are ignorant of anything beyond sight and sound, then we can think of ourselves as innocent and non-responsible for what happens. But if it is shown that we *do* know, then we are responsible, and have to look within for the motivations and the early reasons for wanting to do what we do. And that means a shocking discovery of various elements of the painful roots of our decisions to identify and set up our tiny boundaries.

Thus, the very reason ESP remains subconscious, for the most part, is because the information is so very unacceptable. Secretly we are full of fears, insecurities, propitiations, and dependencies which are in contradiction to our pretentious self-images. To say "yes" to our ESP is to threaten our whole identity with its survival values; it threatens the whole world of our illusions, the entire "maya" of our existence and all those pretenses which cover up the survival threats of the past. This can be very frightening. Thus it is that only bits and pieces of ESP data are generally permitted above the conscious level. What, for instance, becomes of the "accident" in which we injured or even killed a "loved one"? By denying ESP we protect ourselves from accepting responsibility for a thousand events which we can safely call "accidents" and for which we are pardoned. We can thus allow hostile and aggressive impulses to interact, to play out, without searching inwardly for the painful truths about the reasons, origins, motivations, secret identities, and self-images.

Beyond these personal reasons there is a second and larger category of reasons for the general denial of ESP. It lies in the fact that most of our social institutions are likewise based on the non-responsible innocent victim myth. Going all the way back to the burning of the witches, it extends to questions of property rights, frauds, liabilities, crimes, justice, and punishment, international relations, treaties, wars, trade restrictions, political parties, elections and the exercise of political power, economic contracts. Thousands of elements of the social fabric are firmly established on an anti-ESP basis. It is not a neutral position, nor based on lack of information. It is blatantly opposed to subconscious realities.

This being the case, what is our position? We are saying that Psycles shows, demonstrates, or proves the universality of ESP. But because the social structure is based on the non-responsible assumptions which are essentially anti-ESP, it would be foolish and naive to propose sweeping changes. What is permissible is for individuals to change. We are allowed to liberate ourselves. We can *use* Psycles to do so. And, step by step, in decades, or in centuries, the institutions will gradually change as appropriate to be in step with however and wherever the people lead.

Now, let's look at some of our raw data.

The Suicide Attempt

On October 10, 1952, between 10:00 and 12:00 on a late Friday evening, there was no reason whatsoever for my good friend Alice to be feeling so depressed. Everything seemed to to be going normally as far as her life was concerned. Her two grown children were out in the world taking care of themselves, and she herself had a responsible position as a nurse-therapist which was interesting and rewarding. And yet, suddenly, she was overwhelmed by a deep, apathetic, "sunk" feeling. Being aware of the Psycles phenomenon, she scanned events 36–37 hours back, and three days back, and found nothing to account for her serious emotional upset. About midnight the disturbance went away and she went to sleep feeling quite normal.

About noon the next Sunday (36 hours later) she was again overwhelmed by the same feelings! Only this time they were much worse, bordering on despair of life itself. She flung herself on her bed and sobbed heavily for almost two hours, baffled by the lack of reason for her feelings. Gradually, the acute anxiety and grief subsided, and by about 1:30 P.M. it was gone.

About 5 that afternoon the phone rang. The call was from about 50 miles away.

"Your daughter tried to commit suicide this noon. She's out of danger."

Now, what went on between those two bodies? Obviously there was communication. But what sort of communication? Obviously, subconscious. Alice was "in touch" with her daughter both times: the Friday evening crisis which set the trigger, and the Sunday noon suicide attempt, 36 hours later.

ESP anecdotes can be found in large quantities. But this one has the special significance of embracing both events—the trigger and the suicidal response—one 36-hour Psycle apart.

Some Routine Demonstrations

Out on a Sunday drive, a little boy of about two (who happened to be my son) took a wiener sausage which he had in his hand and proceeded to smear it all over the car window. When asked to quit it, he burst into an indignant screaming and hollering.

Because the reaction was so exaggerated, I made a quick Psycles time-check and discovered that about 36–37 hours earlier his mother had had a similar burst of paranoia and self-pity. So, while driving along and without looking at him, I imagined him saying to himself, I don't *really* need to be like my mom . . . I don't *have* to feel like she did. . . ."

And within that or the next second, he suddenly and abruptly ceased his drama, smiled, and turned his attention to some pleasant activity. The concept had hardly been formulated, and "heard" in his ears, when he changed his behavior completely.

Anyone can do this as a routine proof of ESP. There is no skill involved.

On another occasion, we had just finished the evening meal when we heard the screeching of brakes and a loud crunching collision of a car about a block down the street.

By the time I arrived at the scene, there were already a dozen persons standing around a car that had bashed its front end pretty far in against a telephone pole. It was a touring car, with top down, and in the front seat were the male driver and his female companion. She was screaming hysterically, nonstop. But she was not obviously hurt.

Without even looking at her, I *imagined* her voice, within her own head, saying, "All *this* on account of yesterday morning? Because of what happened Tuesday morning, we *had* to do *this*?

And in less time than it takes to tell, she suddenly terminated her hysterics. In an instant, she was calm and quiet. No one had done anything to her.

Again, anyone can duplicate this. There is nothing magical about it (unless *everything* is magical!). The fact is that there is a total sort of communication which exists, subconsciously, between living bodies. The fact that we don't know exactly what it is, or how it works, is irrelevant.

I am especially pleased with the work of Cleve Backster and for his beautiful demonstration of what he calls "primary perception" between people and plants. He hooked up lie-detector electrodes to broadleaf plants, and discovered that they were highly sensitive to his thoughts, and responded dramatically when their well-being was threatened.

To test their sensitivity to life in general, he set up an

automatic apparatus which, at a random time in the middle of the night, upturned and dumped a bowlful of brine shrimp into a pot of boiling water. Under continuous recording, his plants showed an intense reaction at that precise time. He even had his garage door hooked up so that when he returned home and thought the right thoughts, the signals from his plants would trigger the garage door to open!

If ESP is a function of molecular signal resonance, then there is no reason why plants should not be involved in the whole subconscious global communication because they have the same identical molecules (porphyrins, DNA, proteins, sugars, etc.) that we all have!

The Attic Fall

While visiting one evening in the secluded resort of Inverness, in California, I witnessed an "accident" which in its own way "proved" ESP. It was about 10 P.M. Tuesday, when my hostess suddenly asked her nephew, age 22, who had flown in from Long Beach the day before, if he would mind getting a carton of old photos from the attic.

After placing a six-foot stepladder in position, the nephew climbed up, and with one foot braced against the far wall, lifted the trapdoor with both hands and was giving it a half-turn in order to lower it through the hole to us below. Suddenly, he lost his balance, and nephew, trapdoor, and ladder all came spinning and crashing to the floor in a mishap which left him shaken but unhurt.

"Well!" I began, after we were all comfortably settled back in the living room. "Can you imagine the possibility that you did that on purpose?"

"I fell on purpose? But that was an accident!" He frowned in disbelief.

"Allright. Let's see if we can find the reason. Here it is Tuesday evening. Going back about 37 hours—what were you doing in Long Beach yesterday about 9 A.M?"

"Yesterday morning?"he began reflectively. "I had the usual appointment with my therapist. But this time, he began probing an entirely new area of my subconscious, which scared me. I clammed up, retreated, and told him that I didn't want to get into that whole subject."

"Aha! There it is. The reason *why* you did the fall. From that incident to this fall is a one-to-one, cause and effect relationship, like stimulus-response, showing its purpose."

"You're putting me on!"

"No. No. I'm serious! What I'm suggesting is that your slipping and falling was engineered with precision, and that you needed to do it for reasons. And that the reason can be clearly seen in the incident Monday morning, 37 hours before.

"Look at what happened," I continued. "Note how your therapist was threatening to enter the dark attic of your subconscious mind (symbolically), and how you were afraid of this new unknown, and how you wished to avoid it. It's as if the fall dramatized the danger, and rationalized the avoidance, making it correct to stay out of there."

But where does ESP come in?

If we accept what the biological clock seems to be saying, and recognize that this timing is proof that the fall was not a coincidence, and (aside from the symbolic meaning it clearly shows) was done on purpose, and was intentional, then the whole attic fall reeks of subconscious communication and ESP. The fact is that the whole episode was initiated by the aunt. It was she who suddenly set it up. She had no conscious knowledge of the therapy session the day before, nor her nephew's need to do something about it. But when 10 P.M. arrived, she responded to his unconscious need, and spontaneously provided the avenue for action.

We have scores of Psycles case histories which literally cry out that ESP had to exist as a necessary condition for the events to have unfolded as they did.

The Bee Sting

If I don't have a real fun time on New Year's Eve I feel cheated, robbed or "stung." On this particular occasion everything went wrong, and I felt very annoyed with my wife's intransigence that blocked our going to some party, and as a consequence we remained home and angry.

What I did about it 37 hours later is another one of those stories which demands ESP as a necessary condition. January 2 found us in Palm Springs, touring some art galleries. At about noon we entered a large one, which had an ornamental tree as part of the decor. I did not go around hunting for a bee. And this one was quite invisible from the time I entered the door. I went to the tree and casually put my left hand around the trunk, and on the far side placed my ring finger on top of a bee which immediately proceeded to thrust his stinger into my flesh right beside my wedding band.

It was done precisely and perfectly, with full subconscious knowledge that a bee was there. And the bee was perfectly cooperative with my needs.

The Suicidal Bird

While driving along a suburban road, I noticed out of the corner of my eye a small bird circling about fifty yards ahead of the car. Suddenly, it made a sweeping curve and with perfect aim dove straight into my radiator, killing itself. It had all the appearances of a deliberate suicide.

Freud and others have speculated on the symbolism of flying birds in dreams and often associated it with death wishes. In my case, about 37 hours earlier, there was a specific incident, actually a feeling, which involved a suicidal impulse and the mystique of death. Based on this giveaway timing, then, was there not some form of contact-communication for that bird to become martyr for the needs of my psyche? Or, better, did this little bird have its own reasons for suicide, and then tune in to and come together with my signals?

Animals do commit suicide. Years ago a doctor in Thailand told me of an elephant who committed suicide. It was born with a defective tooth which grew in sideways. The elephant was fed on mash until it was about five years old, but appeared to be in pain much of the time. One day, unable to bear it any longer, it made a deliberate 50-yard charge across a jungle clearing and smashed its skull in the trunk of a large teak tree.

The Psychic Dog and So Forth

A chemist friend of mine in Cincinnati named Wolfgang Mueller was strolling with his large and well-trained German shepherd dog. Without warning, his dog suddenly tore at an angle across the street and bit a total stranger in the leg. After profuse apologies, it turned out that the newcomer to the area was also named Wolfgang Mueller.

What unconscious signals *must* have transpired between man and dog to bring about this totally contrary behavior on the part of the dog?

The Disney movie of the book *The Incredible Journey* illustrates a well-documented case in which a trio of two dogs and a cat, abandoned in western Canada, cross 3,000 miles of continent to find their master's new home in eastern Canada. Turtles, grunion, whales, and salmon migrate through thousands of miles of ocean, always knowing where they're going.

Bird migrations are another evidence of global awareness. One hundred billion birds migrate incredible distances all over the globe, some at great heights, some going 25,000 miles on their round trip, through calm or storm, day or night, moving unerringly to their targets.

For left-brain oriented scientists to seek to pin their travel knowledge on various simple physical cues is rather pathetic, I think, but reflects the general denial of ESP.

A Review of ESP in Some Earlier Cases

In Chapter 1 you may remember the Flaming Hand case, in which a young father burned the same guilty hand with which he had beat his infant baby in the middle of the night. Let's assume from the circadian timing that when he stopped to aid the stalled motorist, it was his *intention* to burn his right hand because of the remorse he instantly felt after spanking the baby. Could this have happened *without* the appropriate awareness and communication (ESP) to bring about the circumstances so rightly, to secure the perfect cooperation, the timing, everything just so? It was a highly complex interaction.

He didn't have to tell the driver, "Look. It's my desire to burn my right hand. Therefore I want you to do such and such." No, on a heart-to-heart basis, subconsciously, the driver knew what to do to bring it about.

In Chapter 2 there was the story about Boswell in London.

Now, if there were no ESP, and the whore were unaware of Boswell's need to scream at her, she could just as easily have asked a more modest price, and reduced it even further, if necessary, to suit the customer's demands, and there would have been no scene. But again I say that ESP was a necessary condition for the event to have happened as it was *meant* to happen, for Boswell to vent his anger against Lady Northumberland.

The same goes for the various car crashes in Chapter 2. In each case there must be that unconscious knowing, that agreement to come together, to fulfill the mutual needs at that precise time.

In Chapter 8 the suicidal Catholic mother was away from her house when the husband came and abducted her son. It was a short trip to the post office that she made, and we would say that ESP was again a necessary condition for that perfect timing for the (subconscious) cooperation to take place. She knew the husband was coming, and subconsciously she wanted him to rid her of the burden of this troublesome son, so she "conveniently" absented herself from the house at the right time for it to happen correctly.

Also in Chapter 8 was the missed plane incident. Subconsciously, the lawyer *knew* when the plane would be leaving, because it was his *intention* to be left behind, because he *wanted* to vent his fury against a Big Mother. The Psycles timing was the giveaway for seeing these motives, and ESP must again be considered as part of the context of the whole incident.

In Chapter 10 was the case of rape. How could these two strangers come together in a strange location without communication? If the 37-hour timing is the critical giveaway which uncovers the motivation and intent, then we must conclude that a man who needed to rape, and a woman who needed to be raped, came together by mutual consent (however subconscious) in a confluential event in which ESP is again a necessary prior condition.

The near robbery, also in chapter 10, is a blatant case of ESP. The symbolic meaning (the near loss of virginity 37 hours earlier) meant that there was more to it than merely a robber and victim coming together. The robber had to know, subconsciously, that he was not *meant* to get away with the valuables, but to go only so far as to *almost* get them. And that's what happened when he was interrupted and jumped out the back window.

In the next chapter we'll be stretching ESP out a little farther, when we start looking into events in which many people are involved. It is all consistent with the Psycles view that ESP is not just a sometime thing, but that all life is in fact marvelously interconnected.

13 Coming Together for Catastrophes

ARE there no limits to self-responsibility? Are there no innocent victims of anything, anywhere?

With Psycles we have uncovered the hidden motives and the self-responsible reasons for many events which have usually been thought of as happening to innocent people. We have seen it in so-called accidents and mishaps, in illnesses, in strange behaviors, in crimes, and even in the ways we choose to die.

But are there not catastrophes involving larger numbers of people who are truly innocent victims of circumstances? What about plane crashes, industrial explosions, prison riots, or epidemics? And what about "natural" disasters such as floods, dam bursts, electric storms, power blackouts, or tornadoes? Are there no innocent victims in any of these conditions?

The Jamaica Bay Disaster

On Thursday, March 1, 1962, a huge American Airlines Boeing 707 jet thundered down the runway at Idlewild Airport, New York, heading for Los Angeles with 95 human beings—all ready to die. It climbed steeply in clear weather, banked sharply to the left, and continued in a tight turn right down into Jamaica Bay, slamming into shallow water with a fiery explosion. All 95 died in what was the worst single plane disaster in U.S. commercial aircraft history to date. Was this intentional?

"No! No!" some would say. "Look at all those marvelous intentions and plans that this or that person had. Somebody was going to get married; somebody else was going to close a great business deal; and yet another was going to visit his grandson for the first time, and so on. There were many apparently happy and important reasons for wanting to live! They didn't *want* to die, surely? All suicides? Each with his own reason? No! It can't be!"

For many years my Psycles case histories have spelled out the incredible omniscience of individuals doing their individual troubles, whether generating accidents, or modifying their bodies and tissues with various illnesses and malfunctions. The conclusion was unavoidable: people really know what they're doing, subconsciously. And in this process there is vast communication, of some sort, which we surveyed in the last chapter.

Having established the precision with which people do their troubles individually, how could I draw a line somewhere and say that up to this point people know everything, but once they are up in the air they are ignorant? Suddenly, new rules apply? That would be illogical and inconsistent.

So with the assumption that all these people *knew* what they were doing, and that their intent was death, the scanning of the news reports about the individuals involved in many cases revealed subtle clues of such intent.

In this particular case I happened to know one of the persons who went down. His marriage was crashing on the rocks. He had almost taken the plunge twice already! Consider this very carefully. Twice previously he had been booked on planes which crashed. But in each of these previous occasions he had conveniently missed them, either by being late or by a shift of schedule. It would seem that he was flirting with death, but was not quite ready—until that March 1, 1962, American Airlines Flight One. His decision was final. He didn't wish to come home . . . to what?

You might grant that some people had a reason, or that some had a hunch or premonition, but still insist on a majority of the rest being "innocent victims." And I would have to insist that all persons have the same subconscious omniscience and that ESP is not just a sometime thing. Heart to heart, gut to gut, ESP is an always thing. It permeates all life, like a grid or network that interconnects.

On that March 1 flight were every kind of person, young and old, rich and poor. It included multimillionaire W. Alton Jones, board chairman of Richfield Oil and former boss of Cities Service, on his way to join Eisenhower for a fishing trip to Mexico. What deep and impossible troubles was he facing, that he should wish to join that suicide plunge?

Donald Ellis, 35, an engineer with North American Aviation, had just attended the funeral of his mother back East. How was he identified, that he should decide to end it all? George S. Harwell, 35, was returning home to a mother and stepfather, but didn't. And Frank Levie, Jr., age 22, ex-marine, was also returning home to a mother and stepfather, but didn't. How were their identities affected by father and mother divorcing, and where was daddy?

Flight 629, Sabotaged in Denver

On November 1, 1955, there was another air tragedy, killing 44 persons. AP reporter Arthur Everett did some sleuthing and wrote an article called "Real Life Drama Is Too Far-Fetched for Movies as 44 Fly to Deaths." His reason for investigating this accident was that it was sabotaged, and most of the passengers were already on board this cross-country flight when the dynamite came on board in Denver. Twenty-three-year-old Jack Graham placed the dynamite in his mother's suitcase, saw her on board, and then hoped to collect $37,000 insurance which he took out on her life.

Here, then, were 43 other passengers, ostensibly innocent victims, all condemned to death by this greedy and vicious young man. But were they really "innocent"? From New York and Chicago, when they boarded Flight 629, did they not already know that Graham was to be their agent in their respective suicides?

Everett lists some of the "unsuspecting" passengers as "a man trying to outrace death, a woman with a strange gripping premonition, a mother taking a baby to a father who had never seen him, a couple whose lives had just become worth a small fortune, a girl engrossed in her approaching marriage, an executive whose signature was still fresh on his will, a physician marking his seventy-second and final birthday," and so on.

History will be made when an investigative reporter uses Psycles for the first time for such an event. It should become a classic alongside the famous *The Bridge of San Luis Rey* by Thornton Wilder.

The Research Laboratory Explosion

On July 3, 1958, at 11 A.M., a fiery explosion ripped through four laboratories of the Research Division of the U.S. Industrial Chemicals Corporation in Cincinnati, Ohio, where I was employed as a research chemist. About twelve persons were burned to various degrees. No deaths.

The question was: did the subconscious needs of many persons all feed into the deliberate (though subconscious) creation of this event?

After exploring some 15 lives involved in this explosion and finding a host of peculiar coincidences of timely troubles which zeroed in on this Wednesday morning, I am quite satisfied that the answer is yes. If one were to take an uncompromising stand on the issue of ESP and deny absolutely the possibility of communication, mind-to-mind, then such a thesis as this would have to be considered ridiculous and absurd. I am suggesting, on the contrary, that many persons agreed to do this explosion to fulfill their personal psychological needs. Some participated vicariously; others went out of their way to participate directly, to burn themselves, or to feel certain emotions.

Significantly, the one person who would have been right in the middle of it called up sick and stayed home that day. Apparently it was not his need, so he wasn't there. A number of other persons, who had no logical reasons for being in that vicinity at all, just happened to be there. Presumably it was their need. In talking with various "victims," I found that the greatest all-pervading theme to this explosion was self-punishment with hell-fire. It occurred 72 hours after Sunday church-time, with interesting 36-hour events Monday evening.

This is not to imply that all persons caught in explosions are religious. Feelings of guilt are universal. Self-punishment is universal. And fire is a great purifier. About two-thirds of those interviewed had gone to various churches that Sunday. The sermon topics may have stirred guilt feelings, and may have even touched on hell-fire. Thus, if this were an event being planned and engineered to fulfill needs—what would be more logical than to find that three motion pictures currently showing in Cincinnati were *Fire Down Below, Man on Fire,* and *Fire Under Her Skin*.

Analyses of large-scale calamities often hint at broader meanings and a wider significance, but usually miss the target because of failure to recognize the subconscious communication involved. Once we begin to realize that people are in full communication subconsciously, and engineer precisely their needs and requirements, then the real flavor of such events comes into view. In spite of our pretense of ignorance, we really do know what we're doing, and many peculiar "coincidences" make sense. As we have said, there is a species of omniscience which permeates our existence.

The Sinking of the *Titanic*

To pursue this view further, we can take any great disaster and, with the assumption that it represented certain psycho-social needs in which people are deeply involved, look around at the surrounding events to get a better understanding of why it happened.

In the case of the *Titanic,* there was a great ballyhoo about its being "unsinkable." In the context of that era, we find a rising conflict between science and religion. Science (with a capital S) was coming on strong as a new religion, holding the promise of a new age of rationalism and diverting a new generation from old beliefs. As I see it, the whole cooperative effort to sink the *Titanic* was required socially and engineered with precision (subconsciously) to dramatize the fallibility of Science and give it a slap in the face as an upstart for being so pretentious. And once the drama started to unfold, thousands of individual lives and their requirements began to fit into place and participate in the catastrophe.

As this giant ship steamed westward across the north Atlantic it had been warned of icebergs in the area south of Newfoundland,

and should have slowed down at the least. But because disaster was the *intent,* these warnings were ignored. And when the *Titanic* struck the iceberg about 10:30 P.M. April 14, 1912, and was sinking, there was a ship a mere ten miles away that could have rushed to the rescue. But, again, because disaster was the *intent,*its wireless operator had conveniently retired for the night, so no SOS messages were received, and it left the scene of one of the greatest ocean disasters of history.

Once you realize that people really are omniscient, you can begin to see a rightness about what people do to bring about that which happens. Clear out the old cobwebs of beliefs which assume non- knowing, and start with the assumption that everyone knows everything. Then, accept the events as beautifully engineered—however catastrophic! Realize that people are omniscient and bring about their destinies with magnificent precision.

Realize also that people can change their destinies! With insight. More of that in Chapter 16.

What about Natural Calamities?

Massive power blackouts struck New York City on November 9, 1965 and on July 13, 1977. Could these have psycho-social significance? It happens that the first followed shortly after President Eisenhower had a stroke. That has all the earmarks of symbolic meaning. Big Daddy having a stroke is much like a blackout of national significance.

It took 13 hours to restore power the first time, and 25 hours the second. Both these timings are delightfully circadian and hence point to psycho-social meaning. But are we suggesting that the lightning that struck the transformer stations was directed by the will of a great number of people? And likewise, would cyclones and tornadoes strike only those areas, farms, and homes that wish to be destroyed?

I would not go so far as to say that we personally control weather conditions, storms, and lightning discharges, although there are some strange "coincidences" in such matters. Lightning bolts do strike transformers frequently.

I personally prefer to think that natural calamities can occur in accord with various geophysical parameters, without human direction, but that what we *do* about them individually is what is self-responsible. In other words, with our subconscious omniscience of how events are unfolding, the point is how do we *use* it to fulfill our homeostatic needs.

In a dam burst, for instance, I would assume that our omniscience includes full subconscious awareness of the earth's giving way days ahead, and if it is not our need and wish to be a victim, we will spontaneously have good reason to go away for a while and not be there when it happens.

And in the case of the person I read about who was struck by lightning the second time, I don't see him as controlling the lightning bolt and directing it onto himself. No. What I see is a man who wished to be struck, felt into the sky, and knew where the energies might focus, and then moved to the right place under a tree to get blasted.

Of course, he didn't consciously want to be on the receiving end of a lightning bolt, but these kinds of stories add weight to those who seem to prefer the idea of a capricious God who can zap anyone, anytime, if they don't mind their P's and Q's.

This includes . . .

Sniping Massacres

One of the more horrendous incidents of such apparently meaningless encounters with death occurred in Austin, Texas, on August 1, 1966, when Charles Whitman, ensconced on the thirtieth floor of the Texas University Tower, proceeded to shoot 44 people.

We are not so concerned with his motivation. He may have been identified with a father whom he hated with a passion, but press accounts did not reveal any good clues to the trigger incident, presuming one existed. But what about all the students, professors, casual visitors, and businessmen who were shot? Bodies were strewn for blocks in every direction. Did the 13 who were shot fatally "just happen" to be passing in view of his telescopic sights, without any reason for wanting to be shot? And did the 31 others only want to be wounded?

In many of our self-injury case histories, we have found meaningful precision in their execution; this has pointed to specific motivations, and thence to the self-responsibility which lies so unnoticed everywhere. We therefore question exactly where and how each victim was hit. Among Whitman's "victims" were the newsboy on a bicycle who took a bullet in his buttocks (significant?) and the pregnant woman who took a bullet which merely wounded her in the stomach while killing her (unwanted?) fetus. From all the rest of our 2,000 case histories, we cannot dismiss the idea that each of these hits was purposive and confluential.

To paraphrase Khalil Gibran, Whitman was the unwitting agent of all these victims' wills in the fulfillment of their individual needs. Or, put another way, all these victims *used* Whitman's murderous hostility to accomplish their own requirements. This means cooperation and confluence. A confluence is where two rivers flow together to become one.

The Highway 101 Sniper

In another incident north of Los Angeles, a sniper killed three and wounded ten in a spree of shooting at motorists speeding along Highway 101. A high school junior, Michael Clark, took a high-powered rifle, selected a good vantage point, and on an early Sunday morning, April 25, 1965, proceeded to pick off his victims at random.

But just how random were these "victims"?

William Reida was on his way to a funeral in Portland, Oregon, with his wife and four children. He was hit in the neck, but survived, whereas one of his sons, age five, was killed. What was there about the funeral that they couldn't face, and how was the son identified?

Joel W. Kocab, an aircraft engineer, 28 years a bachelor, is our giveaway clue. Because Friday evening, a Psycle earlier, he had announced his engagement to be married. Was he trapped? with deep conflicts? such that death was the only way out? Because about 36–37 hours later, early Sunday morning, he was dead, shot by a total stranger, far away from home. That Psycles timing is our evidence that it was willed, and that Clark was to be the instrument of his own suicide. Tuned to his agent, he came to the right place and the right time to get what he wanted.

Prison Riots, Etc.

If I were a prison warden, and a massive riot broke out, I would get on the bullhorn and, no matter how wild the idea may seem, start blaring out the events which occurred 37 hours and three days before, to direct the prisoners' attention to the prior times. There are various ways that it could be done without sounding stupid, and from my 28 years of Psycles research experience, I am prepared to predict that a sudden quieting of emotions might occur to defuse the whole event. This implies, of course, that the warden would have to talk openly of the problems existing in the trigger incident uncovered by the Psycles backtracking, and promise to be judicious and sympathetic in its treatment.

Besides prison riots there are many other social and institutional situations in which Psycles can be used. Many cases have been cited in this book to show that directing attention to the prior trigger incident can result in an immediate alleviation of symptoms. So don't hesitate to try it! You have nothing to lose but some old beliefs that may be shopworn.

14 How Psycles Shows Up in Both Primitive Societies and Modern Science

If what we're saying is essentially true, and all over the world people are responding to certain stresses with the same circadian time-lags (Psycles), wouldn't you expect to find this manifesting in social institutions?

Among the many books on primitive practices and taboos, there is a classic called *The Golden Bough* by Sir James G. Frazer. It is full of fascinating stories of what people do in different societies. We were particularly interested in scanning it for examples of how various societies dealt with that critical three-day reaction time, or homeostatic lag-time, and are pleased to be able to reproduce the following selections, taken directly from the book.

The Three-Day Taboos

With regard to the Creek Indians and kindred tribes, we are told that they 'will not cohabit with women while they are out at war; they religiously abstain from every kind of intercourse even with their own wives, for the space of three days and nights before they go to war.'

Among the Angoni, to the north of Zambesi, warriors who have slain foes on an expedition smear their bodies and faces with ashes, hang garments of their victims on their persons. This costume they wear for three days after their return.

When a Nandi of East Africa has killed a member of another tribe, he paints one side of his body, spear, and sword red, and the other side white. For four days after the slaughter he is considered unclean and may not go home.

In the Pelew Islands, when the men return from a warlike expedition in which they have taken life . . . are shut up in a large council house and become tabooed. . . . After three days they go together to bathe as near as possible to the spot where the man was killed.

Among the Inuit or Esquimaux of the Bering Strait, the dead bodies of various animals must be treated very carefully by the hunter who obtains them. Hence the Unalit hunter who has had a hand in the killing of a white whale . . . is not allowed to do any work for the next four days. At the same time no one in the village may use any sharp or pointed instrument.

Similarly among the Aleuts of Alaska the hunter who had struck a whale with a charmed spear would not throw again, but returned at once to his home, and separated himself from his people in a hut specially constructed for the purpose, where he stayed for three days without food or drink, and without touching or looking upon a woman.

Once more the soul of the grim polar bear is offended if the taboos which concern him are not observed. His soul tarries for three days near the spot where it left his body, and during these days the Esquimaux are particularly careful to conform rigidly to the laws of the taboo.

Among the Hottentots, when a man has killed a lion, leopard, elephant, or rhinoceros, he is esteemed a great hero, but he has to remain at home quite idle for three days, during which his wife may not come near him.

Similarly the Lapps deem it the height of glory to kill a bear, which they consider the king of beasts. Nevertheless, all the men who take part in the slaughter are regarded as unclean, and must live by themselves for three days in a hut or tent made especially for them, where they cut up and cook the bear's carcass.

In Madras it is considered a great sin to kill a cobra. When this has happened, the people generally burn the body of the serpent just as they burn the bodies of human beings. The murderer deems himself polluted for three days.

Among the Esquimaux of the Bering Strait, during the day on which a person dies in a village no one is permitted to work and relatives must perform no labor during the three following days. It is especially forbidden during this period to cut with any edged instruments, such as a knife or an axe; and the use of pointed instruments, like needles or dodkins, is also forbidden.

In Samoa, those who attended the deceased were most careful not to handle food, and for several days were fed by others as if they were helpless infants.

Thus in Fiji the taboo for handling a dead chief lasted from one to ten months according to his rank, but for the commoner it lasted not more than four days.

Similarly in the Motu tribe in New Guinea a man is tabooed generally for three days after handling a corpse, and while the taboo lasts he may not touch food with his hands.

In Annam mourners are not allowed to marry or to have any sexual relations whatever during the three days after the death . . . and during these days they are further forbidden to chew betel, to drink alcohol, and to eat flesh.

In some Dyak tribes men returning from an expedition in which they have taken human heads are obliged to keep to themselves and to abstain from a variety of things for several days; they may not touch iron nor eat salt or fish with bones, and they may have no intercourse with women.

Again among the southern Massim of British New Guinea a warrior who has taken a prisoner or slain a man remains secluded in his house for six days. During the first three he may eat only roasted food and has to cook for himself. Then he bathes and blackens his face for the remainder of the time.

Among the Kayans of Borneo men who have been absent on a long journey are secluded for four days in a small hut specially made for the purpose before they are allowed to reenter their own homes.

Maybe our western civilizations should reestablish some of these lost taboos. If, for instance, it were taboo to use knives or guns or drive cars or planes for a period of three days following a death in the family or job loss or marital break-up or blow-up of a love affair, I predict there would be a sharp drop in "accidental" injuries and other traumatic events, especially if that three-day period were spent in "retreats" designed for that purpose.

Psycles in the Biological Research Literature

The modern research scientist goes to his laboratory and usually works between 8 A.M. and 5 P.M., goes home, eats, plays, sleeps, and then returns to his laboratory for another day. Almost never is he/she in the laboratory at either 12.5 or 37 hours after the start of the test. These times are in the middle of that or the next night. There may be frequent observations during the first 12 hours, but then a break until 24 hours, 48 hours, and 72 hours is typical.

Years ago I read in the journal *Science* about an experiment in which a culture of living cells was blasted with a lethal dose of X-radiation and then photographed with a special infrared camera at various times to see the results. There were no visible changes at 2 hours, 4 hours, 8 hours, 12 hours, or 24 hours. But when they returned to the laboratory on the third day and took a picture at 48 hours, lo! all the nucleoprotein in the cells' nuclei had disintegrated. I immediately wrote them to suggest that they repeat the experiment and take pictures at 34, 36, and 38 hours, to check whether the events had occurred in a narrow time-frame of 36–37 hours. They did not respond.

Now, if our facts about Psycles time-lags are essentially true and permeate all life, then the evidence should be found throughout the experimental literature in the life sciences. Because, in biological research there are thousands of artificial stresses imposed on various life forms and which are followed by a variety of responses at various times later. The Psycles nature of these delayed reactions, I believe, has been missed by the researchers in circadian rhythms because, first of all, the responses are singular rather than rhythmic, and secondly, because of the confused and inadequate reporting of the actual time-lags involved.

If, for example, an observation is negative at 24 hours, but then positive at 48 hours, the actual time-lags are lost forever in such phrases as "within 24–48 hours," "48 hours later," "at 48 hours ," "about 1–2 days later," "about 2 days later," "within 2 days," "on the third day," etc., none of which is helpful in calling attention to circadian time-lags of stress-reactions. As a result, none of this massive data was recognized as significant.

In Appendix A, I have included a chart of some 16 examples of this type of information. They were selected from about 350 which I have on file. Six of them report time-lags of exactly 36 hours. We assume some rare and heroic people behind those experiments. The rest are either the "between 24 and 48 hours later" or the "72 hours later" time-lags.

When root tips are exposed to X rays, the first abnormal cell division mitoses don't show up until 36 hours later; when nasal washings are transferred from "cold" patients to the noses of healthy volunteers, the first "cold" symptoms do not appear until "between 24 and 48 hours later"; when bone marrow is grafted into kidney or tongue tissue, calcification doesn't start until 72 hours later, and so on.

There is massive data of this sort which we believe supports Psycles.

15 Is It Genetics or Is It Identity Problems?

In this chapter we come to the "bottom line" of what we really are. We confront head-on two opposing views of life which have been fighting it out through history.

Are we gods or are we robots?

Some of the basic issues which the two views seek to explain are the age-old problems of human suffering, animal instincts, creative genius, twins, IQ, homosexuality, as well as what, basically, determines behavior.

In this book we have been saying that Psycles shows that we are truly marvelous creatures. Though we may suffer, we do it with magnificent precision, being creative, purposive, and fully self-responsible, with vast subconscious awareness bordering on the omniscient.

The opposing view, held by the majority of the BioSciences community and spearheaded by the sociobiologists, is that we're not. They say everything is by chance. Free will and self-determinism are myths. We are environmentally determined robots. For them, the bottom line is genetics.

They claim that the above problems can best be explained genetically, that you are what your genes determine, and that you are bound and limited by the molecular inheritance which sets not only your structure and basic functions, but your social and psychological behavior as well.

We say that's nonsense. It's not genetics except for basic structure and basic functions. In the majority of social behavior, we're looking at the choice of models and the carrying out of those patterns with which the person has chosen to identify.

They say all suffering is determined by genetic inheritance; it is pure chance what happens to innocent victims in accordance with statistical laws.

We say No, No! They are treating bodies as if they had no heads! Actually every cell of our bodies is under our control and modified in accordance with motives, wishes, and desires; it's kids deciding to be like mom and dad—for reasons—and choosing to duplicate their behaviors, traits, and their sufferings.

They explain away all friendship, gratitude, trust, sympathy, and self-sacrificing altruism, whether among people or among animals, as merely deriving from "reciprocal altruism." They say that the only thing that matters is the survival of the individual genes. The genes direct us in a way that maximizes their own survival and our bodies are only temporary mobile homes for the genetic material.

We say that "genes" are only molecules made up of chains of atoms. Since when did molecules behave like people and have minds to make choices?

These opposing views are illustrated in the accompanying sketch against a background of a magnetic field. What we are saying is that the Robot View of the BioSciences represents the bottom half of the picture only, whereas the Psycles view encompasses the entire magnetic continuum.

Since genetics is so crucial, let's explore the whole subject, and see what new insight Psycles has to offer.

What Is Genetics?

All living organisms are composed of tissues which are composed of living cells. Each cell has a nucleus which encloses enormously long molecules of nucleoprotein hereditary material. Called DNA, it is in turn composed of four basic units (symbolized by A, C, G, and T) which are strung together like beads on a chain.

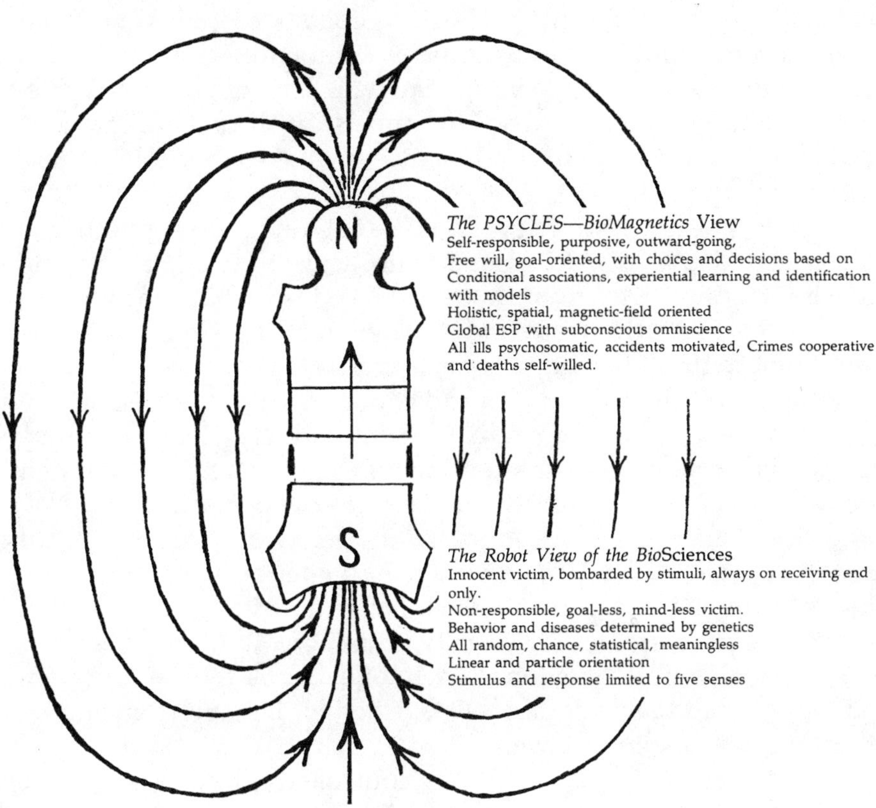

Fig. 11

The sequential order of the units of this template determines the specific structures of the rest of the cell (proteins, fats, carbohydrates, etc.) and hence whether the life form will turn out to be an ant, a rose, or a camel.

Although geneticists act as though their science were clean, clear, and definitive, the truth of the matter is that they're playing with their hands in a "black box" without knowing exactly what they're doing. The structure of DNA is still a mystery, and the Watson-Crick double-helix model turns out to be a fiction, though useful (*Chemistry*, October 1947, p. 22) (*Science News* 116:420). And there are scores of serious contradictions despite scientists' being able to make new forms of life by fusing new DNA combinations in bacteria.

Years ago, for example, while they were were first beginning to piece together what they call the "genetic code," it was stated that if there should turn out to be *one* ambiguity the whole idea would have to be scrapped immediately. Life would be impossible; you might get ants instead of camels! Just one amino acid difference in hemoglobin, for instance, makes the terrible disease sickle cell anemia.

Well, guess what? They *did* find that ambiguity—one triplet UUU coded for two different amino acids, phenylalanine and leucine (*Scientific American,* March 1963, p. 89). But that horrendous fact was neatly swept under a rug, and the code goes merrily on. Something is terribly wrong there!

Some of the many contradictions within the science of genetics are listed in Appendix H. Though highly technical, they should be brought to your attention just to let you know that their data-base is confused and they are knee-deep in molasses.

Their claim that all our living behavior, including our psychological natures, is governed by genetics is actually quite logical, strangely, in its own way! It has to do with the assumption that living processes are essentially "chemical," and that a lengthy series of chemical reactions, each one zapped by a different enzyme catalyst, underlies all our living behavior, including thinking and feeling as well.

If it were true that enzymes controlled all behavior, since enzymes are protein molecules, and since proteins are determined by DNA, then it follows logically that all our thoughts and other behavior would also be controlled by DNA. Therefore all is genetic.

But this is not true.

Appendix H describes a massive upheaval on the horizon for the chemical view of life. There is another revolution brewing which says that life is not just "chemical" but, in fact, electromagnetic at the finest molecular events, with magnetic fields at the very core of life itself.

This new revolution in the secret of life should cut genetics down to size, limiting it to structure and basic functions. Thoughts and behavior, on the other hand, are complex functions which involve magnetic fields, waves, signals, vibrations and resonance— all of which are outside the field of chemistry, and hence genetics!

A Brief Review of the Psycles Case

In this book we took, first of all, the simple "accidents" and showed that they are not so accidental after all, but engineered with marvelous precision, for specific purposes. This was the first big breakthrough for Psycles.

Every clear case of a Psycles time-lag preceding an "accidental mishap" is evidence, *ipso facto,* that that behavior was not genetic, but a function of choice, motivated, purposive, etc. You may remember the following:

The Flaming Hand in Chapter 1,
The Grass Whip, the Near-Crashes, the Car Troubles and Collisions in Chapter 2,
The Honeymooners, the Lathe Bolt, and the Bike Detour in Chapter 2.

Each of these cases illustrates two important features: the precision engineering involved, and the perfection of the awareness (subconscious omniscience) that was needed for it to have been brought about as precisely as it was. Genetics is OK for the structure of the body and its basic functions, but not for all we do.

Then, in Chapter 3 we took a side trip into the body-mind system as a whole to show how we are unified, holistically, as "psychosomatic" beings. We reviewed some of the reasons why we cling to the non-responsible, "victim" view. Contrary to medical practice, we said that bodies have heads! We reviewed the physiology of emotions and how the "physical" and "mental" are like two sides of one coin. Molecules have vibrations, and put out signals. You remember the seesaw reciprocal functions, and the evidence that disease and healing are all holistic events, involving the entire system as an oscillating, para-crystalline network of cells.

In the next chapters 4–7, we showed how Psycles unravels a huge spectrum of "diseases" to show how and why they are generated, on purpose, with precision. This included everything from sneezes to heart attacks, hypoglycemia, epilepsy, and cancer—all precision distortions which are motivated and not "accidental" at all. Again, every clear case of a Psycles time-lag preceding a specific "disease" turning on is evidence, *ipso facto,* that that one is not genetic, but a function of motivated choice, intent and purposive. This included the following:

Various internal distortions including heart troubles in Chapter 4, Hypoglycemia in Chapter 5, Epilepsy in Chapter 6, Breast Cancer and Warts in Chapter 7.

In the next four chapters, 8–11, we showed the same precision perfection behind the various behavioral distortions, from crazy feelings and depressions to missings, losings, and even suicidal deaths, and from crimes to nightmares. Again, nothing "just happens." All these kinds of behavior are logical in their conditional ways, and done with reason and purpose.

Then, in Chapter 12, we came to the crucial subject of ESP, and how bodies are in tune with each other all over the place, with a vast subconscious omniscience which is global. We say crucial, because without it much of our case becomes absurd. ESP is perfectly logical on the basis of resonance between protein solenoids and their signal vibrations.

Every Psycles case history which also demonstrates subconscious communication is *ipso facto* evidence against the "Know-nothing" sociobiologists and their claim that we're all idiot robots. The following cases all prove ESP:

The Flaming Hand in Chapter 1,
Boswell in London, the Car Crashes, the Near-Burglary, Bike Detour, Lost Car Keys, President Ford's Collision in Chapter 2,
The Missed Plane and the Suicidal Mother in Chapter 8,
The Shooting of Mr. Booth, the Sniper Victims in Chapter 9, in
The Rape, Mugging and Police Brutality in Chapter 10,
The Suicide Attempt, the Suicidal Mother, the Attic Fall, Brian's Wiener, the Hysterical Motorist, and the Bee Sting in Chapter 12.

All of these cases, and many more, attest to the vast subconscious awareness which shows our self-responsible motives and intents to do what we do as free agents. They also show how conditional and identified we are.

Thus our case is based on the unity of the body-mind, its

psychosomatic nature, and the subconscious communication (ESP) which permeates all life. This is a self-responsible view of life, in which people choose to identify in certain ways and then generate their unique behaviors.

Natural Selection and the Lion's Pee

Inasmuch as natural selection is the cornerstone of Darwinian theory in biology, let's look at some of the contradictions therein. I shall touch on just a few to show how that position is vulnerable.

Now, natural selection is a quantitative process, a step-by-step process, in which capabilities that are of assistance in survival persist over generations of time, to become traits, naturally selected. There are a vast number of animal behaviors which are not only highly intelligent, but which reveal prescience, careful calculations, and highly specific goals which are desired and which may be pleasurable but have no significance in terms of pure survival.

I was visiting at the Broadmoor Hotel Zoo in Colorado Springs some years ago. In the lion's cage the female was lying in one corner, while a large male paced back and forth rhythmically, paying little attention to the human visitors. As I was departing the area, I heard a group of people coming around the large circular structure, approaching the lion's cage from the far side. There were two couples in their fifties. In the lead were two women whose voices were shrill and nasal, proclaiming personalities which were vulgar and lacking in compassion or feeling for the animals.

The huge, maned lion timed it perfectly as the four came into range. The instant they stepped in front of his cage, he made his own turn, and with back to them he peed a tremendous blast of urine right into their faces, and continued his pacing with complete nonchalance as they screamed in anguish.

Even fish do fun things. One famous fish can jet a drop of water into the air to strike an insect, shooting it down. This is not a survival matter, either, because careful studies have shown that its diet is overwhelmingly internal to the water, and only a tiny portion consists of insects shot down. This is in absolute contradiction to natural selection and survival based on the advantage, gained little by little, of such capabilities. Can you imagine one accidental hiccup which brought down an insect leading to this fish's considerably refined skill—selected for survival reasons, over generations?

The bone and muscle structure of the woodpecker's head which permits it to drill into trees with rapid hammer blows without scrambling its brains is another contradiction to natural selection.

And the penguin's pouch is another. An egg in the Arctic would freeze in an instant, but the penguin scoops it into a warm pouch, which is another complex structure that could never have been developed little by little, over generations, by natural selection.

Throughout the animal and insect world there are thousands of fascinating features of life which show ESP communication, and a precision engineering of structures and functions which are motivated in accordance with needs and desires. They are not a function of random, statistical, "just happen to aid survival" chance events. They are not "naturally selected" over generations because of promoting "survival of the fittest." Humans are not the only animals that have heads. Other animals control their cells too.

Nevertheless, the thrust of the entire BioSciences community is toward the bottomless pit of the random, meaningless, happenstance theory of natural selection, the cornerstone of that greatest cop-out for avoiding truth—genetics.

Can We Disprove Genetics by Proving Identity?

Much of what is assumed to be genetic we believe to be kids choosing to be like mom or dad, assuming a subconscious identification, setting up resonance with a basic image-model, and then duplicating much of that behavior.

The simplest proof of this would be examples of persons who, after many years in a specific identity, changed to a new identity with a whole new set of psychosomatic indicators. Unfortunately, this is rare, because few people ever unlock their central image, which is so buttressed by life-or-death decisions and reinforced with survival significance.

But it does occur, now and then. Usually, it occurs at death's door or "the end of the line" of one identity, with a shift to a new one. Examples can be found among those who have gone to the Lourdes shrine in France, for example, having played out their role right to its final stages, and then "miraculously" switched to a new health and a new image. There are some radical transformations which come out of religious conversions.

But the key to whether these represent true identity changes lies in the physical manifestations. The person must start doing a new set of psychosomatic distortions, a new pattern of symptoms, diseases, accidents or physical modifications. What we have said is that all our physiological distortions and cellular changes are under our control and done with precision, in accordance with models. Therefore, if a person switches to a new identity at the deepest level, it will result in changes in his physical structure and functions depending on what the new model is.

Experimentally, we might take a family in which most of the children are locked into some "disease" which is considered to be genetic because it "runs in the family." Divide them in two parts, leaving one alone, and give intensive psychotherapy to the others, adopted individually into new families with new cultural structures. This might include "block-busting" psycho-active drugs that may help dislodge the parental identifications with the deepest insight.

Some years ago there was a research program at the large state hospital in Costa Mesa, California, in which the mushroom drug Psilocybin was administered to autistic children. Their ages ranging from three to twenty, these children spent most of their lives in a totally involuted state, often curled up fetally, completely oblivious to the outside world. There was no way you could communicate with them. Up until the age of two or three, for instance, they might have been "normal," but then turned inward, to shut out the world.

A strange thing happened with Psilocybin. When the drug was ingested, and during the period of several hours when it was active and before it was metabolized and broken down by body metabolism, the children almost all "woke up" and were communicative and playful. But when the drug "wore off" or was metabolized, they returned to their womb-tomb, again oblivious to the external environment. Unfortunately, during these "open" times, there was no effort to query these children on what triggered the withdrawal, and why they made the decision to do so. The program was suddenly terminated because of adverse publicity during the height of the LSD paranoia in the late sixties. I'm not aware that this type of program has ever been tried on any other "genetic" problems.

Down at the bottom of the list of 1,500 diseases that are called "genetic," there are of course some genuine cases. In the case of sickle cell anemia, there is one amino acid difference in the protein part of the hemoglobin molecule. That is genetic. I suspect, however, that many of the others are merely identities that "run in the family" and could theoretically be changed with the proper motivations. In the Downe's syndrome (or what was known as mongoloid idiots), in which the genetic roots are not all that clearcut, could it be possible that a prenatal life-threatening trauma associated with a screamed "You goddamned idiot!" might be deconditioned?

One case of identity shift that I know was the son of close family friends. He was about five, was grossly fat and lethargic, with oily skin, many pimples, and repulsive behavior. My son Brian was the same age. Shortly after one of our visits, this boy announced to his parents that he was changing his name to Brian, and within an incredibly short time, was totally transformed—slim, trim, healthy, and bright, and remained so. It was a deep and permanent shift of basic identity, demonstrating choice rather than genetics. The human body-mind system is marvelously flexible and malleable. The genes controlling his proteins, fats, sugars, and enzymes did not say whether his body would be repulsive or handsome. This one boy disproves the Robot View.

Similarly, I knew a couple who were married for eighteen years before they were divorced. During this entire time, and previous to their marriage as well, the wife was slim and trim. After the divorce and a remarriage which brought to the wife some new insecurities and anxieties, including elements of violence from her new husband, her overall physical structure changed dramatically. She became obese.

She had two sisters who were large-bodied. Her mother was moderately matronly, but not obese. The idea that it was her "genes" that determined that she would finally, after 55 years, suddenly become fat is nonsense. She deliberately (though subconsciously) modified her body and adopted a new self-image. It was an act which was intentional and motivated. Not genetic.

We tend to act the way we were acted to: the child who was brutalized tends to brutalize his children; a child who was loved, tends to be loving, etc. This is a function of identifying with models, not genetics.

I was listening to a psychiatrist, Dr. Daniel Hunt, of the University of Washington, give a lecture on male-female differences. Throughout his talk, I noticed, he kept inferring that babies were un-learned when they were born, coming out with a clean slate, so to speak, as if the nine months in the womb involved no learning, associating, conditioning, identifying, responding, no feelings, no awareness.

And I asked him, "Throughout, you seem to assume that the baby at birth has no psycho-social history behind him?" He agreed. That was his position.

"But," I pursued, "an amoeba or an isolated muscle cell can learn, feel, and associate; they can be given a conditioned reflex. Is a child at birth with trillions of integrated cells less of a life than an amoeba? The nine months *in utero* represents many signals, many pains, many sounds; a child obviously hears and feels in the womb."

His reply was that there were no animal studies which would tend to confirm that position, therefore he stuck to his.

Fantastic, I thought. What blind ignorance! But then, maybe when he was born, his doctor may have treated him like a chunk of meat. Now he is a doctor, and he in turn is passing on that insensitive attitude toward life.

Again, this is not genetic; it is learned, identified, experiential.

Young Martyrs

Children everywhere, by the millions, are acting out roles relating specifically to the deepest subconscious problems of their fathers and mothers. Tuned into the gut realities of the parent, and knowing their motivation toward some deadly goal, the child spontaneously and unconsciously begins to try to rescue the would-be victim, to offset the drive and save that life.

One example is a young man who developed such outstanding swimming capabilities that he missed the Olympics by a half-second in the tryouts. But all his capabilities failed to save his mother. Deep within her was the purpose to drown (through whatever identification I don't know) and some years later she did so, internally. She drowned herself by filling her lungs with fluid.

From recognition of our subconscious omniscience, I conclude that the son was deeply aware of his mother's intentions, and tried to offset it by his own development.

Deep within my own subconscious, only partially remembered, there lies one of the most horrendous events of my life. I was almost swept overboard by a giant wave, a *tsunami*, caused by an ocean-floor earthquake in the Philippine Sea. The wave came up out of nowhere and swept over the ship on which we were traveling from Thailand back to the States on sabbatical.

At that time I was just over two years old, just taking my first steps on deck unattended, with the awareness that I was about to be displaced as "baby" by a younger sister in my mother's womb. When the wave caught me, I was tempted to allow my self-annihilation and go overboard into the bottomless ocean, but a Belgian passenger grabbed my arm around a vent-pipe and I was saved. There was that instant of a life-or-death decision. Did I really want to die? Did my parents want me to live? I faced the possibility of extinction and, I am sure, communicated with my father and mother with the question—Shall I go? And apparently we all agreed—No, stick around. So I was grabbed and saved.

Now, guess what I recently received as a gift from my son! A framed portrait of himself, surfing magnificently within the shadow of a gigantic 15–20-foot wave off the northern coast of Hawaii. And there he was, cutting through the water, with it cresting high over his head, a fantastic mastery of a threat to life within my subconscious!

In recognition of the ubiquitous efforts of children everywhere trying to save their parents from their deepest subconscious death-threats, we have massive public efforts to save them. Polio is one of the big fields of effort, children being paralyzed (for whom?), being innoculated by the public health authorities to stop it. The battles against the death penalty, cancer drives, MS drives—all part of massive efforts to say, "We are deeply grateful!"

All of this may sound quite nonsensical, unless and until you realize that we are not the ignorant and innocent pawns of circumstances we pretend to be. As we have said, there is a species of omniscience which permeates all life. Psycles reveals it.

The Homosexual as an Identity Problem

From the homosexual who emphatically declares that he has no problem to the psychiatrist who would electro-shock the homosexual out of his horrible "disease," we have dozens of theories and beliefs about homosexuality—none of which are entirely true. We wouldn't enter into this turbulent subject if we didn't feel that Psycles sheds some clear new light on it. Because here again Psycles reveals with pinpoint precision some of the circumstances, relationships, and decisions which underlie why we act the way we do.

Homosexuality is no different from the rest of our "troubles," inasmuch as it is a conditional function. It certainly is not genetic. We are not born that way. Nor does it relate to body chemistry, or what we eat. In essence, it stems from crisis decisions made early in life, which resulted in specific identifications, with models, for how to be and what to do.

Let's go back to fundamentals.

As newborn infants and small children, what we needed most of all was a feeling of physical resonance. Every cell of our bodies was one-half mother and one-half father. Bursting with energy and facing a new environment with many new sensations, we required an in-tune-ment, a feeling of one-ness, acceptance, love, affinity, contact, and warmth from both parents. We needed support and assurance for our growth and survival in a new world. We needed to feel wanted.

But did we get it?

Because mothers and fathers are so mixed up with their own infantile problems, conflicts and mis-emotions, it is only rarely that they can respond with complete acceptance. Often it happens that we may get some love, contact, and affinity from one parent, but not from the other. And this can start a long chain of events and actions to tie us in knots today.

Let's consider the story of Sally.

Her mother had her own problems, worries, conflicts, anxieties. She tried nursing Sally for a week or so, and then started ramming a bottle at her. There was Sally, her whole body wriggling and squirming for love and attention. When she cried, her mother was irritated. Changing diapers was accompanied by feelings of revulsion. Rarely was there the warmth of contact with her mother's body. Lying alone in her crib, her whole body longed for her mother's presence. She got little of it.

On the other hand, Sally's father, though somewhat clumsily, felt more real affinity for her. He wasn't around very much, but when he was, he would pick her up, fondle and play with her.

Sally grew out of diapers, toddled around, crawled and then began walking and doing things. All this time she was like a bright and shining mirror, reflecting subtle patterns which her mother could not stand to see and couldn't accept. Instinctively and spontaneously, like a tuned crystal, Sally manifested subconscious traits of her mother's hidden personality, aspects of her mother's deeper self which had been suppressed, as painfully conditioned.

Mother tried to stamp out these unwanted reflections; she disciplined her, spanked, scolded, reprimanded, made her stand in the corner, or sent her off to her room. All Sally wanted was her mother's love and affection, but it was impossible, because mother couldn't accept Sally any more than she could accept her own hidden, secret subconscious self. Although mother was actually rejecting only the unacceptable reflections, Sally felt personally rejected. Thus the roots of the whole rejection syndrome, underlying homosexuality.

Now, in the course of doing many little things, there were times when, as though by accident, she got her mother's attention and approval. She drew a little picture which her mother thought was cute. So she drew more pictures. It was like a shiny coin which bought affinity and acceptance. She experimented with other forms of action and behavior, and repeated those which seemed to work.

But she couldn't help reflecting her mother's secret self—the subconscious patterns—and for this she was punished, scolded, and lectured. The shiny coins did not always buy what they were supposed to buy—the love that her mother never could unreservedly give her.

One thing, deeper than the rest, Sally tuned into and realized as a very fundamental reality. That was the elementary biological union which occurred between males and females. She always *knew* in her insides what went on between them in the next room. And she *felt* the difference between the two parental contacts.

Now we come to the supreme crisis in everyone's life. It is the focus of Janov's Primal Scream therapy, though he describes it somewhat differently.

Sally had to make a life-or-death decision. Up to this point she was essentially unconditional, unidentified, a "free soul." But in that state, she was a thorn in her mother's side, a threat so severe to her mother's self-image and equilibrium that if something didn't happen— Sally would have to go!

Children's lives are terminated in many ways, and we all agree to be blind to the murder-suicide which it is, because we don't see a deeper perfection about it all. It can be an "accidental" suffocation, or a nutritional disease; sometimes the "sudden infant death syndrome" involving pulmonary edema and hemorrhage, or even cancer.

But whatever the method, it is a "gestalt" agreement, a confluence, a cooperative function to terminate a situation which is intolerable. That is death for the child.

Now for the alternative: the survival decision to *be* like daddy!

If Sally were to identify with her father, relationships could be comfortable for all concerned. This would mean physical survival—but spiritual death. She would have to agree to die spiritually as a "free soul," and become locked into a resonant state of identification which was tuned and *like* father. It is an agreement to become very limited, circumscribed and "normal." It is the transition from the unconditional to the conditional.

This describes the central secret of all human troubles, and the ultimate target of all systems and paths to self-liberation.

It is a physical event. One para-crystalline structure becomes *resonant* with another—like a piano string to a tuning fork—and some central intersect gets locked permanently onto that other oscillating signal structure, and they vibrate together. In resonance.

Having made the decision to identify, Sally now found that her mother could *relate* to her. It had its reward. She was no longer threatened with annihilation. She purchased the right to live by selling her soul, becoming conditional, normal, and human—just like the rest of us.

From then on, in many subtle ways (sometimes not so subtle), Sally began acting like her father. And as the years passed, these physical and mental attributes, personality traits, mannerisms, and interests grew and developed. Becoming more masculine, she decided to become a career woman, and compete with men. Her voice dropped a little. She wore business suits, larger pieces of jewelry, adopted a hair-do which was smart, severe, mannish. She became decisive, independent, and developed business acumen.

All of this, remember, was in accord with *her* particular model which was her chosen identity.

Then Sally married. But it didn't last long. She wore the pants in the family, while her husband played a more passive role. She wasn't conscious of her hostility toward men in general, but there was conflict there, in the reversal of sexual roles, and violence, and a break-up. She became harder and tougher, competed successfully with men, seducing them, and then emasculating them. And all the time, deep within, she was struggling to win the affection and approval of other women—trying to make good the rejection she still felt from her mother, which was like an insatiable hunger.

She even fancied that she hated her mother. But when her father died and her mother moved in with her, she found, surprisingly enough, that she got along very well with her mother in her new masculine role. In fact, her mother encouraged her in various subtle ways to even stronger positions. If there wasn't love, at least there was respect and approval.

Sally had few close women friends, and they were usually quite feminine in their personalities. But though close, it made her feel uncomfortable to have them touch her affectionately. Why? In deep conflict with her fundamental drive to play her identified role was what society felt about homosexuality.

And then a strange thing happened. She began to develop rheumatism just like her father's. After various therapies and treatments failed, she finally consulted a psychotherapist.

That began, for her, a whole new adventure in self-discovery.

She began to unravel the threads of her life. Little by little she discovered the chain of events which led to her present-time situation. She experienced again the terrible pain of that intolerable loneliness and rejection from her mother. She cried bitterly in recalling the many times she had tried to get her mother's attention and love, and failed. She pieced together the various ways she had sought to rectify this, and followed the patterns of action she had taken.

And, one day, after many adventures of self-insight and revelations of feeling, she began to drop her masculine role. Her voice became softer, her body more shapely and more relaxed, with less tension and more grace. Her friends remarked at the changes occurring in her. Her rheumatism vanished. She quit her executive job, and is thinking of marriage again. And that's as far as we can carry her story. Because it just happened the other day.

We wouldn't tell this story if it were unique all in itself. But because it illustrates some common chains of causation involving similar motivations, confluences, and agreements as are found in most lives, it can be instructive.

More importantly, where does Psycles fit into all of this?

Sally's deconditioning, or process of unraveling, was immensely speeded up by the use of Psycles. Throughout therapy, she kept a detailed diary which became a fountain of information. Every little slip, bump, accidental mishap, mis-emotion or upset became the opening wedge for precision insight into her secret-self.

Let's summarize the essentials of this story as it might apply to a man. We'll call him Jim.

Jim requires love and contact from both parents. He gets some from mother, but father is the one who punishes and rejects. Longing for his father's love, he feels rejection, isolation, loneliness.

He discovers certain ways of acting and doing things which happen to bring him some attention from father. He repeats these, but one by one, they fail. Father is still the cold and hard taskmaster who cannot accept the mirrored-image reflected by his son.

With his consciousness of the man-woman relationship, and in the face of possible annihilation, he agrees, in a crisis situation, to give up his unidentified state and be *like* his mother. He dies spiritually, and becomes a normal identified human being. In a subtle dramatization of a female role he is handsome, almost pretty; he wears rings, has a way of flicking his ashes which is graceful; his voice is soft and a little high, and so on.

Likewise, he tries the arts, drama, music, science, and may drive to the heights of creative genius—all in the continuous, though unconscious, attempt to appeal to men and mankind. Please, if you can't accept me, at least accept these, the creative products of my genius, with which I am identified.

Jim may marry a more masculine-type woman. They may or may not make a go of it, being in confluence on a reversal of roles. He may also develop some of his mother's psychosomatic ailments, and act in many ways like her. He has hostility toward women, though he doesn't realize it, and may dramatize the attributes of a Don Juan in order to conquer and punish these rivals for the affection of Man. At one level against this feminine role there is the socially conditioned horror of homosexuality. He fears bodily contact with men, though fundamentally driving for their attention, esteem, and affection. And, against this conflict, he does many things on the side to continually "prove" to himself that he really is a "Man." In this byplay he may go to great lengths to dramatize the pursuit of women, especially beautiful and desirable women, but if they pursue him, he may run.

To solve Jim's homosexual problem, he has to go back to and contact the pain and grief, the longing for his father's love. He must follow the chain of events resulting from the deep feeling of rejection.

There are many Jims and Sallys in this world, identified with a parent of the opposite sex, forced into conflicting roles, with much confusion and duplication of the corresponding psychosomatic problems.

This is the essence of the homosexual problem, as it troubles a majority of the population.

Evidence from Identical Twins

Some fascinating studies are underway with identical twins who were separated shortly after birth, adopted into different family cultures and locations, and then reunited decades later.

There are some incredible duplications of health and accident histories, occupations and interests, pets and children, and so forth. Some of the "coincidences" are bizarre.

Take the "Jim twins" as reported in *Science* 207:1323: "They were adopted as infants into separate working-class families in Ohio and never met until adulthood. Both liked math and did not like spelling in school. Both had law enforcement training and worked part-time as deputy sheriffs. Both vacationed in Florida, both drove Chevrolets. Both had dogs named Toy. Both married and divorced women named Linda and had second marriages with women named Betty. They named their sons James Allan and James Alan. Both like mechanical drawing and carpentry. They have almost identical drinking and smoking patterns. Both chew their fingernails down to the nubs.

"But what investigators thought 'astounding' was their similar medical histories. In addition to having hemorrhoids and identical pulse and blood pressure and sleep patterns, both had inexplicably put on 10 pounds at the same time in their lives. What really gets the researchers is that both suffer from 'mixed headache syndrome'—a combination of tension headache and migraine. The onset occurred in both at the age of 18. They have these late-afternoon headaches with the same frequency and same degrees of disability, and the two used the same terms to describe the pain.

"The twins also have their differences. One wears his hair over his forehead, the other has it slicked back with sideburns. One expresses himself better orally, the other in writing. But although the emotional environments in which they were brought up were different, the profiles on their psychological inventories were much alike."

In another set of twins meeting for the first time, "the manicured hands of each bore seven rings! Each also wore two bracelets on one wrist and a watch and a bracelet on the other."

By definition, identical twins have identical genes, so the sociobiologist who claims everything is genetics is delighted for the most part, but slightly disturbed by the frivolous and totally social duplications, such as the seven rings and the pets' names. But, he takes a deep breath, and is prepared to go "all the way" with genetics.

We say "Nonsense!"

What's happening between the twins, separated however far, is that they are in continuous communication. And we are sure that that is the only true explanation for their marrying girls with the same names, having the same number of children, and naming them the same, and so on. There is communication! Between those body-minds there was a contact awareness to which the researchers were totally blind!

Of equal importance are the differences. If all behavior were determined by genetics, their genes being the same, there should be *no* differences between them. So, as the study proceeds, and more and more identical twins that were separated at infancy are found and studied, the research team should tabulate every single difference with great care. It will most certainly bear out that though their molecular structures are the same, what they *do* with their structure is wide open; they can choose to be quite different in their personalities and behaviors in their different environments—if they wish.

The Identity Problems of Some of Our Cases

Throughout this book we have reported cases which illustrate self-responsibility by revealing motivation and ESP communication behind various troubles. Now we would like to recall a number of them because they are also specific for identity. Each of these cases which follow are unique evidence for people acting out identified roles—rather than following the dictates of genes.

In the Grass Whip in Chapter 2, my self injury was motivated by an urge to reinforce a father-image, a good-father identification, and the sociobiologists would be hard-put to attribute to genetics the original wound which my father received from a stingray.

In the Bike Detour, also in Chapter 2, we have the basic homosexual problem coming out of an identity conflict. The course of a coed's sexual path was being diverted by the more aggressive female toward the alternative identity.

In Chapter 4, an infant was vomiting for, and in place of, the father, who felt severe rejection but suppressed the tension. Also in Chapter 4, the Burst Ear Drum registered an effort against a threat to a strong macho identity.

In Chapter 7 we have two cancer cases, both suicidal because of deep identities: for Taft, the failure of a role, and for Connie the inherent contradictions of her identity.

In Chapter 8 the carpenter was deeply identified with Jesus, and 37 hours after seeing the movie *The Ruling Class* with a psychotic Jesus model, he became psychologically disoriented. The case of the suicidal abduction hinges on the fact that she was strongly identified with a Catholic mother-image, and the loss of her son was tantamount to an impossible contradiction. The Schizophrenic Artist showed a deep unconscious identity with a crazy father, which was done to establish contact with mother. None of these were genetic. They were motivated identifications. And the specific reactions to specific trigger incidents as revealed by Psycles shows the underlying identifications, as well as motives.

The murder (suicide) of Mr. Boothe in Chapter 9 was obviously a father-identification, beyond being a "Junior." His death-intent could not be genetic because their ages were so different, and his father was not shot by a cop. Here, again, we have motive, which is anathema to the sociobiologists.

The Suicide Attempt in Chapter 12 showed a straight mother-daughter identification where the mother was "tuned" to both the trigger incident and the daughter's suicide attempt, 37 hours later. It was ESP, not genetics.

The "victims" of the Highway 101 Sniper were of two contrasting types: the son going to his father's funeral (just like Mr. Boothe?) and the other having a possible homosexual conflict (the Aged Bachelor) with no way out but death. Motives and intentions again.

We have several new cases which illustrate identifications, not genetics.

The Dorm Fire

At 6 A.M. the fire alarm brought some 50 sleepy psychologists out of their beds and into the courtyard of a University of California dormitory in Berkeley. There was a fire in one of the rooms, which happened when a cigarette was left on a coffee cup and saucer while the room occupant went to the bathroom. While she was out, it burned down sufficiently to tip its balance and fall into a wastepaper basket below. The paper caught fire, and when the room filled with smoke, the alarm went off rousing the rest of the building.

We know the fire was set deliberately and engineered with precision—though subconsciously—because it happened just 37 hours after a seminar two evenings earlier (5 P.M.). The subject was androgyny. The woman was secretly identified with her father and experiencing some sexual conflicts in her sexual role. At this workshop, the lesbian "fires" were roused, and ergo—the smoldering fire! Can anyone call this genetic?

The Three Fingers

A busy and successful contractor, I noticed, had parts of three fingers missing on his right hand. I asked him, "Did your father have a full set of fingers on that hand?"

He looked startled, and stammered, "Well . . . as a matter of fact, he had only three fingers too!"

Ha! Here he was, cutting off his fingers, one by one, to be just like his daddy! Now, if grandfather and great-grandfather were born with only three fingers, there might be some genetic factor involved. But not here! This was a son identifying with his father, and then deliberately (though quite "accidentally") duplicating the image. Identity. Conditional motives. Choice. Free will. Self-responsible.

That's the Psycles view of life.

The Mystery of the Missing Three

At the age of five and a half, my son Brian had a serious problem. Reports from the preschool program in which he was enrolled suggested a strange defect. Maybe he had suffered brain damage at birth.

Specifically, the problem was that he was unable to handle the number three. He seemed perfectly capable of counting one and two, or from four on. But the number three was blocked out. Possibly a genetic defect?

At home, I tried all kinds of ruses and schemes to get him to recognize the number three, but to no avail. With a series of objects, such as marbles or coins, for him to count, he would invariably skip that number. "One, two, four, five, six . . . " There would be a slight hesitation where "three" should be, and sometimes a slurring "hnh" followed instantly by the "four." When three of us were present, I would say, "I'm number one, mommy's number two, and you're number three." And that would not disturb him. But he was unable to do the same thing. If pressed too hard, he became very agitated, and would cry.

I thought long and hard about this problem. Other than this problem he was a healthy child. Could there be some traumatic conditioning? Some painful memory? Some bad association? Because, without the number three, he would have insurmountable problems.

One day it suddenly flashed!

When he was about a year and a half, his crib-bed was in a small room just off the top of the stairs. Downstairs in the living room, watching TV, was a new addition to our family, a 12-year-old stepson, who had recently come from Arizona. And on numerous occasions he tested me out, to see the limits of the father-control relationship.

Nine o'clock was his bedtime. "OK, son, time for bed." No response. "Turn off the TV. Come on, now, it's bedtime." Still no response. "I'll give you *three* to be up those stairs. One . . . two . . . "

At the sound of "three" the air was charged with electric tension as he raced up the stairs, just inches ahead of me. I don't know how many times this happened. To him it was a game, a challenge. He was getting "to" me with his refusal to go automatically to bed at the specified time, without instigating borderline violence.

Now, up at the top of the stairs was a semiconscious recipient of this drama. Again and again, the number "three" was the signal for a near-violent confrontation between father and son. It was a painful memory.

So now, years later, there was this unofficial diagnosis of brain damage. Nonsense!

How did we solve the problem?

The next evening, bedtime, I sat with him and recounted the stories of the original incidents. "Remember how . . . ?" I would say, assuming his memory of the events (however unconscious), "and how you used to be sleeping . . . and heard us running up the stairs . . . " and I would laugh about it, recounting it, shouting "three" for the chase.

And then I would kiss him goodnight, with a kiss number "one" on the left eyelid, a kiss number "two" on the right eyelid, and a big juicy kiss on the center of his forehead with "three," and he would sink blissfully into his pillow, and be off to sleep.

Within about a week, there was no longer a problem. Three was OK. And I breathed a sigh of relief.

The next time you hear a diagnosis of "genetics" or brain "dysfunction"—remember this story. Those theories have gone unchallenged long enough.

The Head of John the Baptist

Back in the early sixties, when Timothy Leary was spreading the messianic "Turn on, Tune in, Drop out" by means of the soon-to-be-banned psychoactive chemical LSD, there was a young high school dropout named John Griggs who had a series of religious experiences thereby. These events in turn fired him into forming the "Brotherhood of Eternal Love" whose main purpose was the fanning of the psychedelic revolution.

Every Wednesday evening, the leaders of smaller groups all over southern California met secretly to plan strategy for the weekend. Every Saturday evening and early Sunday morning would find the gathering of this strange underground at some remote spot—a mountain stream, desert valley, or ocean shore—in numbers up to over 500 persons. Silently, at a given signal, they would ingest their LSD, lie down, hold hands in large circles, and await their unconditional experiences that came with the dissolving of their ego-structures.

Week after week, many thousands of young persons experienced religious ecstasy and profound psychological insights in a manner which was so supportive that there were essentially no "bad trips." The hand-holding, the ESP resonance, seemed to have a moderating influence on the panics, dissipating the terror which occasionally erupts when the normal barriers to the subconscious pain of the past are dissolved.

Also, there was a little dog who was everyone's "guru," trotting around, keeping a watchful eye on the individual trips, giving a sniff and a lick to those huffing and puffing with some internal crisis. And when someone might jump up in panic and leave the group, the little dog would follow protectively, lend comfort, and guide him back to the group when the panic subsided.

So much for the background of John Griggs. Through his efforts many back-alley derelicts and even heroin addicts were rescued and transformed to lead productive lives through this strange ritual of group-use of LSD. His religious fervor gained him the nickname of "John the Baptist."

Acting out his identification, he sought his Christ, found him, invited him to their mountain retreat and set him (Timothy Leary) up as the guru of the brotherhood. Having done it, he played out the identity by "losing his head" and dying in a Riverside hospital with tubes pumping out his stomach. I see the whole drama as an identity function.

It was not "genetic" that he chose to identify with John the Baptist. And it was not genetic that he chose to act out the losing of his head and dying. If we but quit that fear of the pain of the past, we can begin to see all around us the illusionary roles that people are playing.

To paraphrase Shakespeare, "the world is a stage, and we are but actors, strutting and fretting our hour upon the stage . . ."

But in the next chapter, we'll focus on what to do to change our acts, to be free.

16 How to Transform Your Life with Psycles

Most of us have, at one time or another, experienced a state of freedom which is quite different from our normal condition.

For some unknown reason, our usual fears fell away, our ideas about ourselves dissolved, our limited boundaries vanished, and for one brief and ecstatic moment, we merged with a wondrous and friendly universe in which almost anything seemed possible. We suddenly belonged in a world that was shining and alive. There was a rightness about everything, and for that short while, we glimpsed our central heritage and birthright as a human being.

It was like coming home.

But then our normal personality closed in to envelop us, and we were once again the blind, stumbling sleepwalker, with fears and pretenses, anxiously acting out a petty role with fixed limits and narrow boundaries.

What was it, and what made it happen? Why didn't it last? Can it be recaptured?

That transcendental experience, that moment of freedom, however brief it may be, has repeated itself endlessly down through the ages. And from that unique glimpse into what we might be, has come the realization of our magnificent potential, a goal which has drawn countless persons into dedicated pursuit.

Impelled by a divine discontent with our normal state, we are lured inexorably into the greatest adventure of life. It is the adventure to reclaim our birthright, with rewards not only in the attainment of the final goal, but in changes all along the way as well.

What Is the Supreme Goal?

For one human being to free his total body-mind from the bonds and limitations of social conditioning, and thus attain to the Unconditioned State of Being—this is how we would describe that state of freedom which is the supreme goal for achievement.

We are not speaking of presidents, kings, captains of industry, commanding generals, statesmen, scientific innovators, artistic geniuses, or any other of the thousands of social achievers.

We are speaking of another category entirely.

All those persons would be considered as sleepwalkers, blind men, and ignorant fools, compared with the Unconditioned individuals. All of them had their ups and downs, emotional turmoils, sicknesses, and accidents; they were frightened, jealous, swayed by hates and fears; all of them were actors, dramatizing a role on the stage—while their knees shook.

We are not speaking of them.

Instead, we are speaking of those relatively few, and usually obscure persons, who took up the challenge to penetrate their true self, to liberate themselves from the bonds of social conditioning. We are speaking of those who conquered, who achieved, who emerged to manifest the Unconditioned State of Being, Unbound.

Down through the ages there have been mystics, sages, saints, and enlightened philosophers, who have formulated methods whereby others could achieve this same state. They called the supreme achievement by many names: salvation, perfect peace, oneness with God, nirvana, heaven, enlightenment, cosmic consciousness, self-realization, sanctity, in tune with the infinite, and many others.

Each of these persons used the language of his particular culture and times to describe his achievement. Because of their different backgrounds, the explanations and the symbols they used often seemed contradictory.

But their truth was One.

The human body-mind can be modified, changed, unbound, and freed of conditional limitations to manifest its own intrinsic perfection. This is what they said.

Socrates said "Know thyself." Plato said "Self-conquest is the greatest of victories." Jesus Christ said, "Be ye perfect. The Kingdom of Heaven is within you." The Buddha, Lao-Tze, Moses, Mohammed, Ramakrishna, St. Augustine, and dozens of others formulated and taught methods of achieving, or unbinding, or whatever we wish to call it. And in yoga there are a half-dozen schools incorporating a variety of approaches and disciplines for this exact same achievement.

The Approaches to Biological Perfection

From the viewpoint of biology, all of these formulations can be considered as methods of deconditioning. This is why we define the supreme goal of all religious systems as the Unconditioned State.

The normal, social human being is from infancy conditioned, and under overwhelming social pressures (involving pain and rejection) forced to identify and be like everyone else. To act and react in the appropriate (and very limited) manner becomes necessary for survival. It is an agreement, a confluence, to become "human." It was a life-or-death decision when it was made early in each life.

To undo this process, to unwind, to "decondition" and free oneself from this extremely limited social structure is not easy. It is nevertheless possible. And there are many methods which have been offered by which it can be accomplished, with effort, determination, courage, and time.

Among the many ways and techniques, there are those appealing to every type of person: some are highly symbolic, and others are very matter of fact; some are quite singular, while others are more group-oriented; some are more energetic, others passive; some techniques are more devotional, others more analytical.

But whatever the approach, underlying all of these hundreds of formulas for achieving the unconditioned state, there are a number of basic principles: self-honesty, full recall of the past, acceptance of responsibility for events and emotional relations with others, the extension of internal body awareness, the acceptance of external realities, the relaxation of body tension, and the extension of the feeling of symmetry.

These are all things which a person must do, by himself, to himself, to liberate himself. All of these processes are interrelated. Affecting a change in any of these categories involves the total body-mind. And pursuing any of them to their ultimate extent can result in the Unconditioned State, in which the tremendous capacities and functions of the human body-mind are free.

The Unconditioned State of Being

When the normal, conditional, social, personality structure finally collapses, the following properties of the liberated state may manifest:

1) perfect health: freedom from physical diseases, neuroses, psychosomatic problems and accidents; a powerful body with enormous stamina and vitality.
2) phenomenal mentality: alertness, memory, wisdom, comprehension, insight.
3) immense perceptiveness, transcending time and space; self-awareness of total body structure and function; awareness of others' deepest thoughts, feelings, and motivations.
4) inner quiescence: relaxed, coordinated, purposive.

This is the intrinsic nature of man, released from his self-imposed, socially conditioned limitations. This defines the individual goal. This is the possible, for those who would seek to discover their true self, unconditioned. This is the target for the great adventurer.

Embarking on the Greatest Adventure

We know we can change ourselves. This is the first step. We know that we don't have to remain forever as we are now. We know that our capabilities as human beings extend far beyond that which we now manifest.

We also know something about illusion. We know that people are full of pretense. We know how we fool ourselves, and how there are often deeper *real* reasons for our activities, which we may not be able to find out without considerable effort. We know there is a veil which hides deeper realities, the subconscious from the conscious.

So we read about saints, sages, and masters, and find out that they achieved this state of being. They were not born that way. And although we may not fully understand their accomplishment, we may nevertheless realize deep within, that we can do likewise.

And without even knowing, for sure, exactly what the ultimate target might be—we make that fateful decision!

The whole world shakes. And strangely wonderful coincidences begin to occur. We have embarked. There is no turning back. Come what may, we begin to seek to modify ourselves, to change ourselves, in a more conscious manner.

It means a fantastic adventure, largely unobserved by the social world outside. It is an inner adventure. An unraveling process. A search that goes deep within, and back in time . . .

> Who and what am I? How did I come to be this way?
> Why did I identify as I did—to structure this self-image?
> How and why did I become so limited, so confined, so out of touch with the cosmic?
> How can I change myself? How can I free myself of these limitations?
> What must I do to unbind myself? to decondition myself?
> I wish to be free to attain perfection. I seek God.

These are the cries which have rung from the hearts of seekers down through the centuries. Whether seasoned travelers or starting the adventure afresh, it is the same. We would free ourselves from the chains with which we have bound ourselves. The Unconditional awaits us.

To help us move toward our goal, we find that there are scores of paths and thousands of techniques from which to choose. Inasmuch as some of us are already involved in systems which are insuring steady progress, we would recommend only the addition of Psycles to augment their systems and boost their effectiveness.

For the rest of us who would welcome being guided with a

complete program, we have made a selection of methods which we feel is balanced and effective. This Psycles Program can be considered as a point of departure from which we may still sample, from time to time, many of the other options.

From wherever we are, we embrace our full self-responsibility and act to change our destinies. A divine omniscience permeates our very being. From within our hearts a vast awareness reaches out to the stars and feels the pulse of eternity.

We are ready! May life never be the same again!

Program Outline

A. Statement of Principles: The New Vision of Life—BioMagnetics

B. Inventories

1. Life History Chart
2. Body Map
3. Family Images and Models
4. Goals
5. The Mind Set

C. Things to Do

1. Disco Meditation in the Morning
2. Nutrition
3. The Psycles Philosophy in the Work Scene
4. Miscellaneous Deconditioning Techniques
5. Two Sets of Exercises
6. Enhancing Sex, Business, and Athletics
7. The Psycles Diary
8. Evening Meditation
9. Association with Fellow Seekers

A. *Statement of Principles:* The New Vision of Life—BioMagnetics.

Surrounding the whole earth is a planetary force-field which guides all the myriad functions of life, penetrating through every living cell, down to its finest molecular events.

This invisible energy-field, whose form is magnetic, not only holds cells together to unify whole organisms, but also unifies the interrelationships between all organisms, coordinating all interactions on a global basis.

Because of this unitary field, we are all personally involved in an invisible link with all life, and through tuned resonances, there is a vast subconscious network of contact-awareness, communicating from heart to heart, in a kind of all-knowingness or omniscience which permeates all life in a harmonious synchrony which is congruent with the magnetic physics of the universe.

The discovery of one heretofore unknown Master-key (Psycles) breaks a hidden code, dissolves certain illusions, and reveals for the first time the intrinsic rightness and perfection of a spectrum of nonoptimum functions such as pain, suffering, accidents, diseases, violent conflicts, bizarre behaviors, showing that they do not contradict the unitive perfection in the harmonious symphony of life.

Suddenly, all these troublesome behaviors fall into place, revealing that (1) underneath all the tension and conflict there is a deeper cooperation and a flowing-together of needs and efforts, and (2) underneath all the painful bodily distortions of disease there is a logical subconscious genesis and a precise carrying-out of intentions, with reason and purpose, in accordance with conditional self-images.

With the breaking of this code and the use of this precision tool, scores of human problems can be unlocked and adversities turned into self-fulfillment for an embarking on a new and revolutionary course toward freedom for disease and suffering.

Psycles is a new and unified way of life, which recognizes our innate divinity and affirms our total self-responsibility within a unitive reality that enfolds our entire existence. It brings fresh understanding to the mysteries of disease and healing.

It is a new guide for self-liberation, a new road map for happiness.

Three features make BioMagnetics unique and different from all other systems, teachings, and philosophies:

(1) It has a marvelously simple set of *three operational principles* which cover the whole of life:
  a) What we ARE is conditional— hence the secret of all suffering.
  b) What we can DO is decondition—hence the secret of all systems of liberation, self-help therapies, etc.
  c) What we can BE is Unconditional—hence the secret of ultimate human potentials, the liberated.

(2) It is founded on *three scientific discoveries,* far ahead of our times.
  a) Torroidal magnetics at the central core of life itself.
  b) The reciprocal functions underlying whole-body integration.
  c) The stress-reaction time-lags (Psycles) underlying all suffering.

(3) It has three special rewards in *life enhancement.*
  a) Our divinity is proved, restoring life to the center.
  b) Our self-confidence is renewed through knowing how to solve any problem.
  c) Life is guaranteed to be exciting, full of new insights, new directions, new health.

B. *Inventories*

To change, to transform, to *BE* something different and better, means the need to be more aware of where we've *BEEN,* what we've *DONE,* and where we *ARE.* We want to see our lives as a whole up to this moment in time. This includes our life history, our body distortions, our family models, and our goals. The more awareness we have of the whole past, the easier it will be to change.

1. *Life History Chart*

Take clean typing paper, tape the sheets together lengthwise (tape on the back) and fold them like an accordion, so that you can draw a life-line from conception to present-time-plus, using about two inches per year.

Mark the years, such as in our sample, and begin entering the important events. Places you lived, jobs you held, trips you made, important things you did. Illnesses, accidents, operations, losses, big troubles—whatever happened. Marriages, divorces, babies, schools, competitions, winnings, losings, deaths . . .

What you want is that your life should be an open book to yourself. Remember as many things as possible—the significant things— so that you can get a FEEL of your life as a whole. Make it graphic, neat. It can be fun.

If you can't remember certain dates, enter them approximately, in pencil, so at some future time it can be erased and reentered, if and when you pinpoint it. Check old records, letters, whatever you have to gradually fill in this life-line. Make this a picture of your life. Get both the good and the bad, the pleasure and the pain. You can put in red streaks and blue streaks (bad and good) to make them stand out. Make it real. There's no one to kid but yourself!

Life History Chart

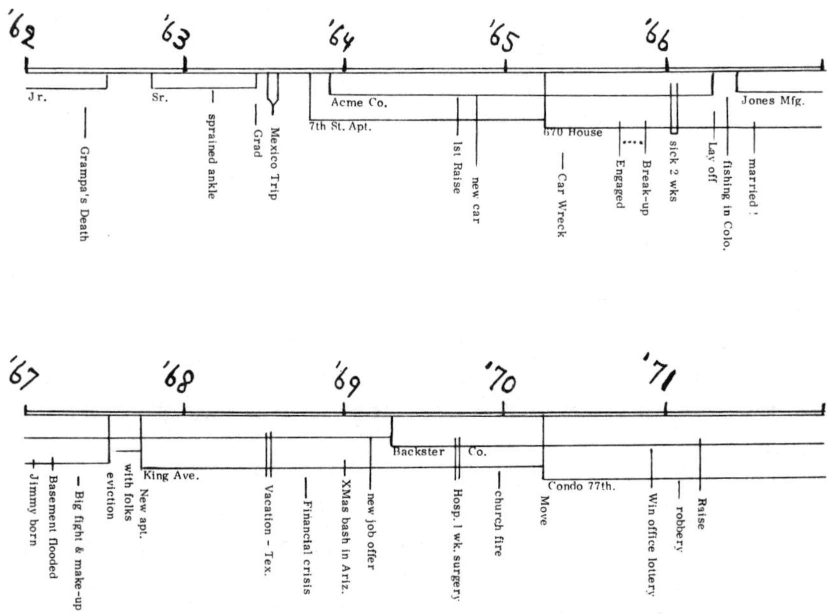

2. *Body Map*

On an outline of the body (you may trace our example if you wish) enter the various nonoptimum conditions and functions, the "distortions" from perfect, that have happened to, or you have done to, your body.

Put a spot on the body that approximates its location, with a line out to an identification of whatever it was. Ask yourself, "What have I done to my body?" Injuries, bumps, bruises, cuts, stitches, aches and pains, itches. At what points is anything "wrong" with the body, now or in the past—diseases, illnesses which you have treated, or were treated for, things which have bothered you, conditions you wish you didn't have: warts, pimples, growths, disfigurations, stiffness, soreness.

Your body may begin to look like a mess, but whose wouldn't? As time goes on and you remember new ones, enter them into the picture. You're not trying to convince anyone of anything. This is strictly for your own purpose, for your increased self-awareness, and with the motive of changing toward something far healthier! So, let it all hang out! Keep it handy, in your therapy folder, to look at now and then, to add to as necessary, and as a reminder of some of the past.

This is part of the landscape you're leaving!

For a Body Map

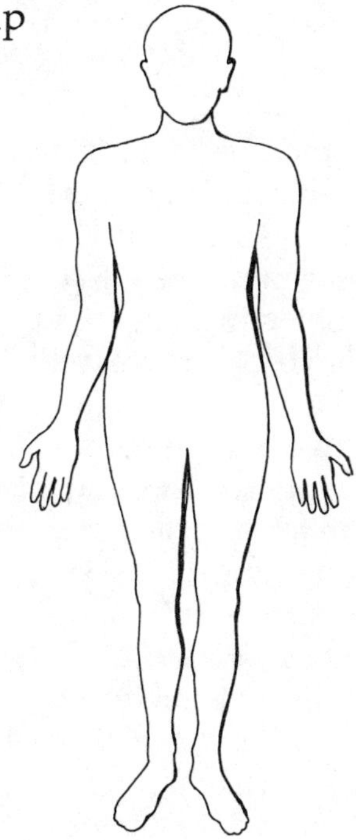

3. *Family Images and Models*

Divide a page with a line down the middle, and then start entering the character, personality traits, likes and dislikes, etc., of your father on one side and your mother on the other. Put down how they acted, how they treated people, how they felt about themselves, their illnesses, accidents, their bodily conditions, how they dressed, what food they liked, how they played, what they found pleasurable, how they suffered, what they feared.

Do the same, on additional sheets, for your grandparents, aunts and uncles, brothers and sisters. What did you like about them? What did you dislike?

The whole family structure, and how you identified, figures deeply in the entire conditional structure you are, and how you set your boundaries and limits. As your whole adventure gets underway, and progresses toward the more unconditional, you will be more aware of how important these patterns have been in establishing the landscape of the life that you wish to move away *FROM*.

4. *Goals*

Start working on a picture-poster of what you really WANT, as goals, and with reasonable luck *should* be able to achieve. (If this is too modest, you can always scale it up later). With both words and pictures (cut out of magazines, etc.) you can paste up a collage which includes your objectives.

Soon, you will have a picture which you will face every morning during your basic program, a picture which will remind you of goals and objectives. It can include pictures of houses, cars, vacation cabin, or other possessions, but also contain depictions of what you want to DO in terms of accomplishments: friends, family structure, educational training, skills, special equipment, facilities, professional honors, successful achievements—whatever you have dreamed of as what you want, and is not entirely out of reach. We recommend that this goals picture be not too outrageous or it will lose its effectiveness. The objectives should be what you can reasonably expect, say within ten years.

Ask yourself, "What do I *really* want?" And picture it. It can always be changed, modified, or upgraded whenever you wish.

5. *The Mind Set*

Impelled by a divine discontent with the way we are now, and realizing that life can be incredibly more marvelous, we must actually make a deep decision to do something about it—to change things. Because, the doing involves certain commitments of time and energy. We have to choose between drifting along as we are, which has a certain familiarity about it even if it is not entirely satisfactory in terms of how life could be. It is so much easier to drift along. A dull, dreary, humdrum life with periodic pain and suffering can be rather comfortable in its own way. At least we know what it is like. Somehow, we are surviving. But to change things? Where might it lead? New dangers? Lose friends? Shake up a whole security structure?

So—courage! Take a deep breath. Know that others have embarked and are glad they did! Know that we can change and that within us are great potentials. The cosmic is at our fingertips. Storm the gates of heaven to seize what is a birthright!

The genuine process of deconditioning can start right in our own home or apartment. If there is the wish to embark on the greatest adventure of life—then decide it! What follows is a balanced and effective program.

C. *Things to Do*

1. *Disco Meditation in the Morning*

The first thing in the morning, every day, is disco meditation. My routine is 40 minutes. That may seem like a long time, but with all that it involves the time goes pretty fast. And it's fun. You may start with shorter periods of, say, 10–15 minutes, providing you do not get some painful kinks or charleyhorses. Don't hurt yourself.

Disco Meditation is essentially jogging in place, with dance variations. The idea is to get as much flexible, rhythmic motion out of your muscles as possible—not too fast or energetic, just a joyful involvement of the whole body. To whatever rhythm is most comfortable, combine the jogging with a light twisting and turning, weaving the shoulders up and down, now and then moving your arms in sweeping circular motions as in a magnetic field.

For a *location* pick a spot which has no squeaky planks or floorboards, nor things which rock and shake and rattle. You don't want to distract yourself or your housemates or neighbors with peculiar rhythmic sounds early in the morning. The space should be such that you can swing your arms freely without hitting anything. Also, you should face a *Mirror* at about six to ten feet so that you can see most of your body. At about eye-level, either on the mirror or to one side, whichever you prefer, tape or tack up your *Goal Picture-Poster.* And on the other side, some whole-body *Anatomy Pictures* (unless you know what your internal structure looks like) should be fixed for reference as you direct your attention to a healthy body-structure.

I usually disco about five minutes for a warm-up (with a robe on if it's chilly) and then stop for the *Body Brushing.* Get yourself a good bristle brush. Mine has a five-inch wooden handle with a four-inch section of brush, in eight rows, like a large bottle-brush. The bristles should be stiff enough to stimulate the skin but not so stiff as to hurt, nor so soft that it doesn't excite the skin.

Whole Body Brushing is a great technique for deconditioning. In a way, it's like whole-body acupuncture. It is therapeutic in that its stimulations are reciprocal to internal tensions or distortions (see Chapter 3). For example, several stiff brushings of the neck (all around, including throat, back, and skull-base) can help turn off (with some Psycles insight into why) an impending sore throat or cold very quickly.

In body brushing, I start with the neck, then up and down my belly, then lumbar (where I need special attention), and then each shoulder, arms, and fingers, then complete torso including up and down the spine, hips, butt, then both legs, up and down, all around, ankles and feet. A once or twice over the body takes another five minutes or so. By this time you should be much more self-aware of your whole body and its boundaries.

Now it's back to *Disco Meditation,* jogging in place, watching yourself in the mirror, dancing, twisting and turning, rhythmically, with some arm motions, finger-snapping and whatever feels good for the maximum involvement of the total body structure and musculature.

Now it's time for the *Countdown to Magnetic Center.* While continuing the disco dancing, say softly to yourself, as in a whisper, "OK, now I want to go to my magnetic center, to my global center, the center of my whole being, my free space, my free and unconditional center. As I count from ten to one, I will go to my magnetic center." Keep dancing, jogging, and now move your arms in sweeping magnetic field patterns, hands together up through the center, and sweeping outward and around, and down and back up through the center, and on, around and around, slowly.

Imagine you are descending in an elevator from the tenth floor, and see the numbers of the floors go upward, one by one, as you count them. "OK. Ten . . . nine . . . eight . . . seven . . . six . . . (see the numbers, one full sweep of the arms with each number) five . . . four . . . three . . . two . . . one . . ." Snap your fingers! "Now! I'm in my magnetic center, and from magnetic center I can command this body to do anything!" Feel the whole world as a giant globe, you're standing on top of it, on top of the world, which is round, and the entire sphere is under you. "I'm in my magnetic center, my free space, my unconditional center."

Now for the specific *affirmations*. Each person must construct his own, in accordance with his needs and goals. But for a start, and as a point of departure, here are some *general affirmations* which will apply to most everyone. You will note that they are strictly positive. There are no negative statements, like saying what you don't want. There's a reason for this. Elsewhere we were listing our troubles, distortions, ills, aches, and pains. But in terms of the specific mechanics of this disco process, make it all positive. We are in a specific state of mind and being in which we are planting seeds, so to speak, for the goals we wish to achieve. We are talking to ourselves, to our deepest center, and we want everything that goes into "magnetic center" to be goal-oriented, and toward the manifestation of the optimum.

"From magnetic center, I can command the body to do anything. By virtue of my innate divinity, as a child of God, I now *claim my birthright!* I claim being slim and trim, healthy and happy, and (my goals). I claim my birthright now! It is my birthright to be perfectly healthy and happy. It is my birthright for my whole body-mind to be healthy and happy. I claim it. It's my birthright. And every time my steps go up and down, I am saying yes . . . yes . . . yes . . . yes . . . to my birthright, slim and trim, healthy and happy, and with (my goals)."

Make your ultimate goals translate into as few words as possible. And as you are discoing, if you flash onto new words which express exactly what you want, stop the dance and write them down right away. Continue the disco.

Now we can go into more detail with respect to the body and

its functions. Look the body up and down as you continue your disco meditation, and claim its perfection! Part by part, organ by organ, system by system, structure by structure.

On some days you may spend much time going over the body. Other days you will make sweeping claims of health for the entire body and leave it at that while you meditate on other things.

"I claim as my birthright a completely healthy and happy body-mind—all my tissues, my skin, hair and nails, my muscles, my bones, tendons, cartilage and connective tissues, my whole nervous system and sense organs, eyes, ears, nose, tongue and taste, my blood-vascular system, lymph system, my whole digestive system, glands and teeth—all healthy and happy."

The word "happy" can include the idea of being healthy, if you say so! After all, if it weren't healthy, it wouldn't be happy, would it? On that basis much of the dance can focus on happy, happy, happy this or that structure. And the word "clean" can mean free of deposits, uric acid crystals, toxins, etc.

"All my joints—shoulders, elbows, wrists, fingers—all happy, clean, relaxed, flexible. Happy joints, happy knees, happy lumbar. My whole neck—clean, happy, flexible. Likewise my vertebrae—all happy, clean, and flexible. It's my birthright! I claim it!

"From magnetic center, I command my body to allow my eyes to see perfectly, and allow my blood pressure to be 120/80 (or whatever), to LET my body fall back to perfection, to be automatically restored to perfection. LET the body feel acceptance of the abundance of God. LET the body be an instrument of God's will. LET the body radiate unconditional LOVE. LET the body manifest its best potential. LET the blood circulate freely to restore, feel, and heal . . . to LET whatever is right to happen." These are all examples of affirmations you may use.

How about worldly success? "By virtue of my innate divinity, I claim my birthright, and from magnetic center I command this body to ALLOW fame and fortune. It is my right to be rich and famous, and to manifest (specific goals)." The word "rich" can mean whatever you wish it to mean. It may mean a multi-millionaire with a huge bank account and vast possessions, or, instead, it may signify having all my material needs satisfied so that I can go anywhere and do whatever is right and appropriate.

While disco-dancing and jogging, you may find *musical tunes* humming automatically through your head. Select a positive one (check their words) and use that while you disco. One tune that I happened on which had just the right rhythm and symbolism was the Navy song "Anchors Aweigh." For me it was a go-go song: "Up-anchor, let go of the past, let go of what you're hanging on to, unhook yourself, let's move, get going, into a new future . . ." Among religious hymns there are some good ones. But choose whatever has the right rhythm and a positive liberation message for you, even a Sousa march!

We end the morning Disco Meditation with a hot and cold *shower*. I let the cold water first hit the back of one shoulder, give it a good go-around, then move across the back to the other shoulder, and then down for a good cold blast to my lower backside. Only then do I turn front-on, and by that time it feels fantastic!

Stepping out of the shower, and with a brisk rubdown, you're guaranteed to be several steps advanced on the Greatest Adventure of Life. It's a good feeling, every day!

Try it!

2. *Nutrition*

Consider healthy food as a part of your program. Although we feel that what's in your head is the most important for life changes, what you eat can materially help or hinder the efforts you are making.

For breakfast (and I consider high-fiber food as commendable) I have a bowl of bran mixed with yogurt, wheat germ, brewers' yeast, rice bran, fruit, molasses or honey—all mixed together with a little extra milk. It tastes fine, while I scan the morning paper, and is highly nutritious.

You might enjoy reading the magazine *Prevention*. It is full of claims of how this or that food, mineral, or vitamin solved this or that problem. Notwithstanding its naive lack of information on psychosomatic realities, it points to good goals and presents some indicators of relations between what we ingest and our physical conditions. It has many recommendations on vitamins, minerals, etc., on which I won't comment.

I do take a gram of vitamin C each morning, in a glass of apple juice. I buy a gallon of unfiltered apple juice, pour a little out, add a half-cup or so of vinegar, and this vinegar-spiked juice is always on hand, for a 50/50 mix with water.

Every now and then I *fast* with just water for either 40 hours or for three and a half days (Psycles times). I have also gone 30 days on water, and it wasn't so difficult after getting over the first few days. If water alone is a problem, that apple juice with vinegar makes it quite easy, as a juice-fast. The benefits of occasional fasting are well documented, and it is well to know, by experience, that you are not so dependent on a strict schedule of food intake. The body can go 60–90 days without food, and benefit thereby. (Do check with your doctor before undertaking any prolonged fast.)

3. *The Psycles Philosophy in the Work Scene*

There is "Magic" in the working day and any job can be an adventure, with continuous fruits in the way of personal growth, if you apply the fundamental principles which we have illustrated through this book: subconscious omniscience, ESP, confluential cooperation, and the micro-precision.

What does this mean in actual practice?

Throughout the working scene it means that you can consider everything that happens as purposive, meaningful, and significant for you personally. It means going through the day with a sense of wonder and anticipation. It means a sense of expectation that things are going to happen which will specifically help you in your unfolding, in your deconditioning, in your self-liberation. It means that the world is alive, and things are happening in a cosmic sense.

It means assuming that everyone knows your deepest secrets and is capable of helping you unlock them, one by one, at whatever speed you choose. It means looking at all other people as your personal gurus, if you wish, ready and willing to help you become free. They know everything. And their pretense of going about their business as if they were impervious to your deepest needs is only an act set by your expectations. They may not know what they're doing, consciously, but underneath this surface innocence, they can be and are, automatically, instruments of your own self-liberation.

If you want people to leave you alone, they will. If you want it otherwise, it will be otherwise. Cooperation is at the heart of nature. The predator and prey are united in a dance which is truly cooperative. The aggressor and victim are likewise, in human societies, in a joint adventure in the total unfolding process.

Nothing is more effective than seeing with the eyes of a child, wide-eyed, open-eyed, wondering, waiting, expecting. The goal you keep in mind is your own self-freeing. And the attitude is one of allowing it to happen, inviting one and all to participate in your liberation, asking for it, inviting it, saying yes, yes, yes to the world and all of its potentials for a wonderful unfolding.

You look around you and see a lot of actors, acting roles, pretending to be this or that, being stiff or suspicious, or wary, or serious, or aggressive, or happy or sad, or terribly burdened, or whatever. It's all an act. Each person is acting out a self-image which he has chosen for its supposed (conditional) survival values, and busily doing his trip, his act. Most of them are acting as if they are victims of many circumstances. But you know better. And though you may give them some of the sympathy they ask for, you are aware that, deep within, this is only their act, and their choice, to feel the way they feel, to do the things they do, to get the cooperation they get, for whatever ends.

Each day you think to yourself—I wonder what it's going to be today, to loosen me up, to edge me along the way, to change my structure just a little, to modify that self-image which I am but which I wish to gradually dissolve. I wonder what it's to be today.

Maybe, suddenly, a great deal of work piles up on you, or bosses are acting up. So just flow with it and realize it's their trip, their problem, their act, that you may lock into—or not. You need not interlock with their tension-systems unless you wish to. And if you do, it may mean that there is special significance in it for you. Ask yourself—ha! why should I now be all shaken up over this thing or situation? Automatically, you check the Psycles time-lags backward, to see what events may have transpired the past several days to bring this about as a need.

Thus—go to work, wherever, and whatever, knowing that all events are correct for your unfolding. Keep these fundamental principles in mind—and life will be a whole new adventure!

4. *Miscellaneous Deconditioning Techniques*

There are literally hundreds of techniques for promoting what we consider the seven primary elements of deconditioning, which is the extension of:

a) full memory, recall
b) full internal body awareness and feeling
c) complete self-honesty
d) awareness of full self-responsibility
e) centeredness and bilateral symmetry
f) relaxation, freedom from tension
g) total self-acceptance

As we mentioned in the introductory portion of this chapter, these properties are theoretically manifest in the Unconditioned State of Being. The functions are all interrelated, so that it would be unlikely to achieve any one of them totally, without at the same time achieving the rest.

What follows is a grab-bag of miscellaneous techniques which can help to promote one or more of these processes.

Extend your field of awareness by being conscious of distant buildings and other landmarks, as you move and act in relationship to greater distances. Feel the whole building, the whole block, the whole city, and then the whole earth-globe. Picture yourself turning on the globe.

Try hyperventilating. Breathe deeply, heavily, rhythmically, to a point where you feel a tingling all over the body. For a breathing exercise, try the 1:4:2 rhythm: breathe in to the count of four, hold your breath to the count of 16, and slowly exhale to the count of eight. Or try panting like a dog. Note how quickly it releases turbulent feelings.

Massage and manipulations help extend body awareness. Mechanical vibrators and foot massages are a great help.

Cross-examine your daydreams. Scan back over a chain of idle thoughts. Trace the chain back to its origins and see how one thing led to another. Practice alert passivity by capturing fleeting thoughts, being aware of the flow, watching them, slowing them down.

Be more assertive; less mousy, inhibited, apologetic. Live more explosively, naturally, from the guts out. Be more spontaneous, demand rights, express emotions, don't hold back feelings.

Practice progressive relaxation, muscle by muscle, part by part, always going beyond that point where you think you are already relaxed. Or, at any time during the day, suddenly note your tensions and drop them. Continue to perform the same functions with minimum effort.

Star-gazing was an ancient relaxation technique. Focus your eyes on a star; note how your eyes tend to drift away and jerk back. Relax them and reach out to a steady focus.

Live your day as if suspended from a rope attached to the top-center of your head. Feel straight and tall, draped and relaxed. Walk tall.

Change your diet. Try new and novel foods and combinations of foods. There are a variety of "cleansing" regimes, and other special diets, such as those especially alkaline, or high in protein.

There are many special tricks and gimmicks for extending body awareness and internal communication. In general, just feel all through the various parts of the body, scanning as with a fine-tooth comb, at any time during the day. Go right to the center of itches, or aches or pains, or tensions. Recognize your responsibility for generating these sensations and efforts, that you are doing them on purpose. Feel deeply and extensively throughout the entire body.

Equilateral ear-wiggling can counteract emotional tensions.

Select any pain, go to its center, intensify it, then reduce it, then intensify it again, and allow it to flow through the entire relaxed body.

Close the eyes and focus on a central point. Put the entire effort of the body into "looking" at that point through just the left eye. Then shift the entire effort, looking at this point through the right eye. Shift back and forth, ending up with the centered "look," both eyes equally balanced.

Stand on your head, do cartwheels, or somersaults. Or try sleeping on a slight incline with your head lower than your feet.

Wear a Hindu caste mark in the center of the forehead. Or stick a piece of plastic tape there. Feel it as you move around, being aware of the body-balance with respect to that point. Attempt to focus on this point either while moving around or while sitting in meditation.

Cry deliberately in uninterrupted seclusion. Reflect on the

occasions of grief in the past, great losses, and contort your face like an infant crying. Clench your fists, double up and let it burst forth—the full horrendous agony of the ages!

Feel heart-centered, with love radiating out energetically.

There are thousands more.

5. *Two Sets of Exercises*

The first set is an ancient set of yoga "setting up" exercises, done twelve times, in the morning, facing the sun.

Breathing is indicated by INhale and EXhale.

The action should be done slowly, methodically, deliberately, with full attention directed toward the whole body, especially to the spine.

This second set is sometimes called the "Dervish Method," as it starts with a spinning motion done clockwise. From II to V each consists of an alternation between two positions. Work yourself up to the optimum of 21 times each per day (which may be split between morning and evening).

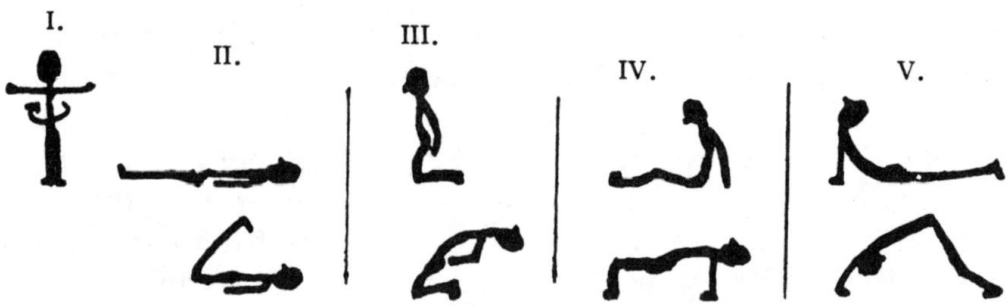

6. *Enhancing Sex, Business, and Athletics*

It may appear excessively simple, but it comes highly recommended.

At about 37 hours before some critical performance, whether it be with a sexual partner, or having to do with a business contract, report or negotiation, or an intensive competition in sports—you can set yourself up for optimum performance by a specific targeting meditation. And if you want to give it the triple-whammy—do the same meditation 75 hours, 37 hours, and again 12 hours before the critical time.

What sort of meditation are we talking about?

A standard sort of procedure would be to sit relaxed, do a progressive relaxation of muscles, count-down to magnetic center, claim your divine birthright, and then declare approximately what will happen at the specified time. Feel the confidence, visualize it taking place, demand the cooperation, and project in your imagination the positive outcome. See it happen.

Now, if what you wish is in real contradiction to your basic self-image and really shouldn't happen, then you would be wasting your time. But if it could happen, and reasonably should happen, then such preparation will stack the cards in your favor and do wonders in terms of your best performance potentials!

7. *The Psycles Diary*

Keeping a diary is a rock-bottom must for Psycles. The time-lags in our reactions to stressful events are at the very core of our new vision of life with all its revolutionary breakthroughs. It is one of the most elementary things to *do* for solving problems standing in the way of that glorious health and freedom we all seek.

So keep a diary. It's simple and easy. Its value for you is tremendous. Even if you do not use it much in the initial stages for analysis of troubles, it is still an important thing to do for its own benefits.

Go to any stationery store and ask for a "Daily Reminder" for the current year. It has a #370 number on the outside, and is manufactured by the Standard Diary Division of the Wilson Jones Company. Almost all stationery stores carry them. They are a handy size, sturdy, bright red, about one inch thick, and measure out as little books, three and a half by six inches.

Start keeping it every day. Every evening at bedtime, make your entries. Of course, you can make entries during the day, too.

But it is most important that this be done every day. And the best time is bedtime when, with a few moments of quiet, you can reflect on the day's events, remembering what happened, and jot them down in chronological order.

We are not recommending the "dear diary" type of essays. All we need to write down is a bare outline, a skeleton of the day's events, each with its approximate time.

As we said, even if we don't do much analyzing at first, it is a good thing to review the day's events each night before retiring. The act, in itself, is very beneficial. Scan the day. Remember what happened. It is healthy to do so. Like clearing the decks, day by day.

It will help you to sleep better. To have the day's events in perspective, to see it as a whole, to recognize and accept what happened, to lay it all out so that you won't be deliberately forgetting or hiding some things. We want full memory. We want our minds to be an open book to ourselves. We want to be open to ourselves. This is deeply healthy.

Jot down the approximate times you got up, and when you went here, went there, did this, did that, through the day. Make special note of any problems, tensions, conflicts, hurts. With this diary you have a permanent record for backtracking any time that you wish to explore the roots to solve any problem. In general, it can be said that the more spontaneous we are in the here-and-now, the healthier it is. The more we deal immediately with our feelings, and don't let up-tight feelings get locked in, the less pay-later problems. So get it down in the diary, and let it all hang out—to yourself.

To help in your analytical work, we include a Psycles Dial Computer on a separate page. You may cut out the central circular dial so that it will rotate with respect to the outer time indicators. Per instructions, and with some additional bits of paper and judicious gluing, it will become functional.

The procedure is simple:

a) What was the approximate time of the "trouble"?
b) Set that time to "mishap."
c) Recall what happened at the various times indicated with arrows. Where was I? What was I doing? How

The Psycles Dial Calculator

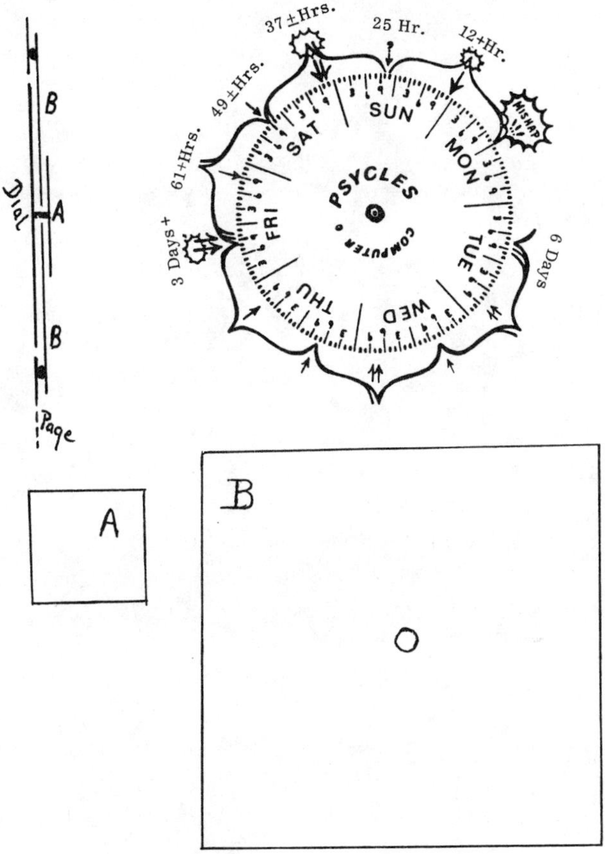

Instructions:

1. Very carefully cut out the circular dial in the center (using sharp razor blade against back-up cardboard).
2. Cut out pieces A and B, including the ⅜″ hole in B.
3. Mark center of back of dial with a dot.
4. Lay dial face down; position B so that its hole is centered on dot.
5. Place a tiny drop of glue on that dot—making sure it does not touch the edges of the B-hole. Lay Piece A on that drop. Do not press down. Let dry.
6. Place this assembly so that the dial fits into its original hole; glue B to the page at several places (under all four corners).

was I feeling? How might they all relate, to give meaning to the "mishap"?

Usually (often at about 37 hours or 75 hours) there will be pinpointed a specific stressful incident of some sort (each person is unique) which you will recognize as: Aha! That's why!

Feel into the incident or incidents to see if you can capture the essence. Reexperience it in your memory. Consider how you might solve the problem it involves. And you can always seek professional help if you wish in solving the specific problem uncovered.

Don't stop at just the 37-hour incident. (Sometimes it may appear blank, not meaningful.) Cover all the arrow-times (the multiples of 12.4 hours) because there may be a sequence of events illuminating the genesis of the problem. And you don't want to miss any additional insight.

Because each person is unique (his whole life history and associations), we cannot anticipate what you may discover for yourself. Nor are the time-lags exact. Unfortunately this relates to the accordionlike nature of the geo-physical events themselves which vary, with a give-or-take in time. So feel around at the approximate times, with a give-or-take of an hour or so. Sometimes you may hit "pay-dirt" right on the button. But the fact that the intervals are not always so exact does not serve to discount the revolutionary nature of the discovery. It's often the difference between knowing and not knowing. This puts us light-years ahead of the professional guessers who are not yet familiar with Psycles!

8. *Evening Meditation*

To cap off the day, sit comfortably, with lights off, and feel a progressive "let go" relaxation of all your muscles from the top of your head to the tips of your toes. Do a ten to one countdown to magnetic center (p. 248), claim your birthright (brief version)—"It is my birthright to be perfectly healthy and happy. I claim it! And to be (your main goal-wish)."

Now add, "With the count of one to five, I will be ready to go to sleep. I will have a good night's sleep, dreaming insight dreams which I will remember, and I will wake when I wish, feeling refreshed, alert, and ready for another wonderful day! OK. One

. . . two . . . three . . . four . . . five!" Snap your fingers. Go to bed (if you're not already in bed), and that's it!

9. *Association with Fellow Seekers*

With the decision to blast through this entangling mass of conditional nonsense, and "go for broke" toward that dream of the centuries—it is well to associate.

In the same way that we described what happens in the work scene, so it is that the same or better can happen with people who are openly committed toward self-insight and change. To deconditioning. Deep within us—all of us—we have that divine omniscience which can prod, support, or nurture our fellow creatures along the way. Unconscious it may be, but it is just as precise in its penetration of the subconscious truths all around us. What we said is that it can work better and more if we consciously invite it. We said people are automatically cooperative. So if our signals are an invitation to mutual help in the journey, there will be that much more interplay and that much faster progress.

Where do we find people to associate with?

There are thousands of groups all over the country, engaged in a variety of systems and methods for liberation. Shun those which are purely entertaining, full of "seat-warmers" who just want to listen to exciting tales of magic power, ego trips, the strange, weird, and mysterious, without doing anything about it. Seek out those groups that are more honestly committed to change, who do certain exercises, practice certain disciplines, and interact meaningfully toward solving personal problems and removing blocks toward their greater self-fulfillment.

Thus—

The paths to internal freedom—the Unconditional State—are many. What we have outlined above are only a selection. You may choose others, which may feel more appropriate to you. Let your heart be your guide. And whatever *your* path, *do* it with enthusiasm. Claim at least a portion of that marvel that is your birthright!

17 Appendixes

A. *The History of The Psycles Discovery and Investigations*

BY the beginning of 1952 I had read in psychological literature of human beings sometimes not reacting to incidents of an emotional nature until two or three days later. Freud mentioned reaction time-lags of several days without further clarification.

I had also come across speculations about periodicity in human events—whether weeks, months, or years, and been aware of three-day incubation periods for some infectious diseases, and read about diurnal variations in some metabolic processes.

But Psycles didn't exist. There was no awareness of a set of predictable time-lags in stress-reactions.

The Original Observation

On Thursday, February 21, 1952, at 5:50 P. M., I rose from the dining table, took two or three normal steps and then, without any apparent cause, my right knee "buckled in" in great pain. Within minutes it was stiff and swollen with intercapsular fluid.

I was puzzled by this senseless event. My knee had been well and strong for years, and at that moment there was no valid "cause" for it to trouble me. Some ten years earlier, I had had an operation on that knee, and had been unconscious while the surgeon took a saw and cut through the patella to expose and clean out a small ulcer on one knob of the femur bone under the cap. I had a ten-inch scar to show for it, but no further troubles.

It just happens that during this time I had been adventuring deeply into psychotherapy, and had recently done an experiment in line with extending my recalls. There, in my diary, was an entry, 4:40 A.M., February 20, in which I had spent a full half-hour in an experimental hypnotic regression back into and through that knee operation. It was my intention to dissolve the unconscious barriers to that memory, and allow that trauma to become integrated in my conscious mind.

The time-lag between the experiment and the reaction was about 37½ hours. Was this time interval just happenstance? Or was it significant, and did it apply to others as well? Was it specific for just this type of event and not to others?

This was the birth of Psycles.

The Search for Case Histories

At first I began checking back through my diaries, some going as far back as 1933 and 1934. I looked for entries such as "sick" or "cold," or describing accidents, and scanned the events several days preceding to see if there were specific incidents of emotional stress or conflict. I found a dozen with similar time intervals between "trigger" and response.

In March of 1952, I expanded my field of inquiry, continuing to observe my own nonoptimum functions against this time calculation, but beginning to use every opportunity that presented itself to learn tactfully about other persons' event-sequences, whenever psychological or psychosomatic troubles were mentioned.

In many cases I landed precisely on incidents of emotional stress or conflict in interpersonal relationships. Often they were clearly associated and meaningful. Sometimes there were "blanks." But in these latter cases I often noticed a catch of breath, a licking of lips, a tension or a startled look which seemed to contradict the protests that "nothing happened." This suggested two possibilities: 1) an incident so charged with emotion that the subject refused to divulge its intimate or sexual nature, or 2) a genuine forgetting which occurred to blot out the memory of pain.

In some cases the time calculation led to a moment when the subject was fast asleep. A few landed on dreams. This suggested the possibility of doubling the period and following the cycles both forward and backward. This proved fruitful.

The First Reports

As a staff member of the Psychological Research Foundation in Phoenix, Arizona, I wrote a brief preliminary report on the basis of a selection of fifteen case histories, adding a minimum of theoretical analysis. This was published by Dr. Gordon Beckstead on October 10, 1952, as Technical Note #5, "Case History Evidence of a Time-Lag of 37½ hours between an Incident of Stress (Restimulating Associative Memories of Early Trauma) and Psychosomatic Responses." One thousand copies were mailed to interested persons including 60 local physicians.

The response from the field was encouraging. Among several dozen letters, we received a series of case histories from a nurse-therapist at the Phillips Medical Clinic in St. Louis, Missouri. These were published November 25 as Supplement 1 to Technical Note #5.

Among our case histories were two which had clearly relevant "trigger events" at exactly 35¼ hours prior to the psychosomatic response. Did these variations in time intervals correlate with lunar cycle variants such as low or high tide intervals, or new and full moon? Apparently not.

The Search of Scientific Literature

An obvious question now demanded an answer: if it were true that human beings respond on a delayed-action basis to emotional/stress incidents which relate to early traumatic associations, would it not also apply to all forms of life, and would not this be reflected in the mass of experimental data from the various life sciences?

I spent the months of April and June 1953 in a literature search in the Colorado College library in Colorado Springs, scanning and extracting from Biological and Biochemical Abstracts. I found that there does, in fact, exist massive documentation. (See Chapter 14.)

A majority of the experimentation, unfortunately, did not follow the metabolic event changes beyond 24 hours, and those which reported for long periods, were often content to make observations at whole-day intervals, thereby obscuring the phenomenon. But 350 examples were extracted of physiological, anatomical, histological, pathological, and psychological responses occurring at significant time-lags following an experimental pain-stress-drug stimulus. (See following table.) Inasmuch as the 36–37 hour interval invariably falls somewhere in the middle of the night following the daytime experiment, it is a wonder that there should be any results reported as occurring 36 hours later. Eleven were found.

Otherwise, most of the delayed-time responses were noted as having occurred "between 24 and 48 hours later," or "within 24–48 hours," or "at 48 hours . . . was observed," and so forth. These were considered significant.

PSYCLES in the Research Literature:

Stress	Response	Time-Lag	Reference
Pregnant rat injected with sex hormone	Induced ovulation; depos. of cholesterol in corp. lutea	36 hours	J. W. Everett—*Endocrin.* 41(5):364-7. 1947.
Starving of rats	Liver begins to lose whole cells, not storage mat'ls.	36 hours	McBride—*J. Biol. Chem.* 147 (2):333-9. 1943.
Root tips of Zea mays exposed to X-Rays	Abnormal mitoses appear	36 hours	Bersa—*Akad. Wiss. Wein Math. Naturwiss. Kl.Sitsung.* Abt. I. 136:383-402. 1927.
Tropical orchid pollen placed in nutrient solution	Germinates	36 hours	Curtis, Duncan—*Amer. Orch Soc. Bul.* 16:594-7, 619. 1947.

Operations involving much tissue injury	Leukocytosis	12-36 hours	Stitt, Clough-*Pract. Bact. Hemat. Parasit.* 1938:340. 1943
Graft of thyroid tissue in mouse ovarian capsule	Evidence of growth	36 hours	Stein, Quimby, Moeller—*Anat. Record* 99:249-64. 1947.
Tobacco mosaic virus injected into mice	Appearance of antibodies	24-48 hours	Libby—*Trans. NY Acad. Sci.* 9:248-257. 1947.
Clucking hens injected with sex hormone stilbene	Clucking stopped	24-48 hours	Prohaszka —*Agrartudomy.* 1:444-8. 1949.
Nasal washings transf. from "cold" patients to volunteers	Onset of "cold" symptoms	24-48 hours	Topping, Atlas—*Science* 106:636-7. 1947
Sonne bacilli placed in rhamnose	Fermentation begins	24-48 hours	Neter—*Bact. Rev.* 6:1-36. 1942
Cholera vibrio placed in egg yolk	Coagulation occurs	24-48 hours	Derkatsch —*Centralbl. Bakt.*1.Abt. Orig. 102:319. 1927
Production of experimental wounds	Onset of fibroplasia	72 hours	Kobak et al, *Surg. Gyn. Obst.* 85:751-6. 1947
Normal subjects placed on subnutrition diet	Basal metabolism drops	72 hours	Husby—*Acta. Med. Scand.* 130:20-5. 1948
Actinomyces griseus placed in growth media	Growth starts	72 hours	Dey—*Sci. & Culture* 12:100-103. 1946.
Bone marrow grafts in kidneys and tongue	Calcification starts	72 hours	Grugni et al—*Arch. Sci. Med.* 91:487-508. 1951.

Experimental thermal burns on dogs	Elevation of serum lipase activity	72 hours	Persky et al —*Am. J. Physiol.* 166:413-415. 1951.

Three Months of Guinea Pig No. 1.

As the most convenient material for a study of the automaticity and cyclical, homeostatic nature of human behavior, I used myself for the most detailed observations. I kept an hour-by-hour diary, noting not only major activities of a routine nature, my comings and goings according to the clock, but also most of the usually ignored and minor "nonoptimal" behavior. This included the tiny aches and pains, twitches, bumps and slips, interactions with people, emotional feelings, conflicts, upsets, ups and downs, all the way to a sudden song or thought popping into my head.

This recording was maintained intensively from May to October of 1953 and charted on some 25–30 feet of chart. The intimacy of the record, the space requirements, and the fact that it involved the private lives of numerous persons, made it necessary to abbreviate and codify the record so that it appears somewhat abstruse to the casual reader. But it is available to researchers and, with its thousands of bits of data, should be subjected to computer analysis in due time. Toward that end I have constructed a set of symbols covering most of the elements of intra- and inter-personal behaviors.

Chasing Down Other Questions

Could the same time-lags be found in detailed diaries written centuries ago? An outline analysis of Boswell's London Journal, 1763, revealed identical timings (see Chapter 2). A similar analysis of the famous diary of Samuel Pepys, written in 1665, is also confirming, though his account is more impersonal and therefore not so rewarding.

Does the phenomenon relate to the Focault Pendulum, or would the same time-lags occur whether at the equator or in the north temperate zone? A detailed analysis of a personal diary from Mexico showed that the timing is the same regardless of distance from the equator.

Is there any straight one-to-one relationships with ocean tides? Apparently not. The intervals between three highs and lows, however, are a variable which spans from 35 to 39 hours, and this may be significant. But the triggers and traumatic responses occur at any and all times between lows and highs.

In one intensive study of a blind man, with long-term monitoring, it was discovered that his body temperature, plasma cortisol, urinary K, C1, Na, Ca, and PO_3, alertness, performance and sleep-wake patterns were all indistinguishable from the lunar day of 24.8 hours (*Science*, 198:423).

In January of 1956 there appeared the first of three articles published by the prestigious Foundation for the Study of Cycles (sponsored by the Smithsonian and Hudson's Bay Company). Their journal *Cycles* carried my research under the title "The 12½ Hour, 37½ Hour and 72 Hour Cycles of Reaction." The impact was minimal on either the economic community (stock market analysts) or the behavioral sciences.

Then, late in the fifties, I came into contact with the rising field of research in biology dealing with the circadian rhythms and realized, for the first time, that what I had chanced on was the circadian half-wave. This provided a logical anchor for my discovery, but still brought no Aha's from the scientific world. Although this field now has a bibliography of over 5,000 items (1980), it is still dominated by the "Robot View" (see Chapter 15). Focused on internal mechanisms rather than a holistic geophysical view, and looking for repetitious events of rhythmic nature rather than response time-lags, it is therefore still quite unaware of Psycles.

The Historic Process

Historically, most conceptual breakthroughs go through a quiet gestation period lasting a score or more years, during which little appears to be happening. If one recognizes the subconscious omniscience shown by Psycles, however, much may be actually happening. There is a preparation going on in the whole mind-grid, a settling-in period in which various contradictory beliefs have to shuffle, adjust, and accommodate, before the new idea emerges into public consciousness. Nothing happens "before its time." When the time is right, it will flow readily and automatically (even to include the public battles and infighting among the protagonists in the social drama). Knowing this, it is easier for us to be patient.

During this interim period I sent scores of queries to book publishers, pressured academia with research proposals, made over 150 article submissions to magazines, had long interviews with important persons, published tracts and monographs, appeared on radio and TV talk shows, in a continuous series of probes in anticipation of that time—when it's right!

A series of five radio dialogues with the famous Joe Pyne, syndicated through 350 radio stations over several years, introduced an estimated 25 million listeners to Psycles, and brought in hundreds of letters, including the one from Cedar Rapids, Iowa, which reported, "Your appearance on the braodcast brought so many calls and letters that the MC took an hour out of his Saturday night for a replay. He had *never* done that before!"

B. The Circadian Rhythm

In Chapter 1 we said just enough about the circadian rhythm to provide a biological basis for Psycles. There is actually a large literature, with a bibliography of over 5,000 items covering a great deal of experimentation on the subject.

Here, as an appendix, we would like to say just a little more about it, partly to explain how it is that we can ignore such a large body of data. It can be summarized as follows:

1) Our environment is rhythmic.
2) Life is electromagnetic, "tuned" to it, and relating to it.
3) Most science has an opposing view, which fears and avoids that concept; it seeks to prove "internal oscillators" rather than cosmic contacts, and hence churns out data which is quite irrelevant and artifactual.
4) There are, however, complex and confusing data which complicate the simple one-to-one correlation between life and geophysical rhythms, and point to new unknowns in the physics of life, light, and gravity.

Our Environment is Rhythmic

Around a spinning sun spins a spinning earth, around which spins a spinning moon. As a result of all these relative motions, the physical state at any position on the earth's surface is rhythmic, oscillatory, periodic. The oceans go up and down. The atmosphere has tidal waves, and every time the moon passes overhead, the very earth itself heaves up and down two and a half feet. We live in a "battery" in which, from ground to sky, there are large voltage differences, and daily rhythms of vertical electrical potential.

It is in this kind of an environment that life is totally immersed.

The rhythmic motions of the ocean tides is of special interest, not as *the* cause of circadian rhythms, but as an indicator of the geophysical events which we believe to be the underlying cause. It would be nice if these rhythms were perfectly regular. The Psycles time-lags might then all be the same. Unfortunately, this is not the case. The following examples, in Figure 12, charting ocean tides and their changing time intervals, show some of the variations which also happen to correspond with the range of variation in Psycles timings.

About half the time, in two separate periods during the month, the highs and lows are fairly even in height.

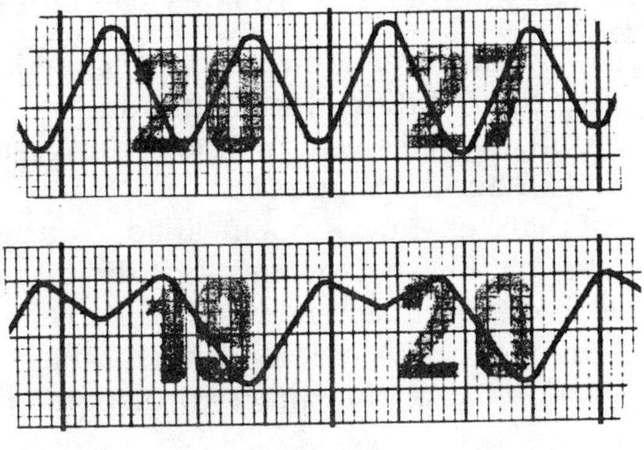

Fig. 12

At the alternate times, in two periods during each month, the highs and lows are more uneven.

The time intervals between lows or highs also vary, in a manner which correlates somewhat with the evenness. Most of the year they vary between twelve and thirteen hours, with an average of 12.4 hours, the lunar half-day. But four times a year, they reach extremes of eleven- and almost fourteen-hour intervals, as follows: The usual pattern:

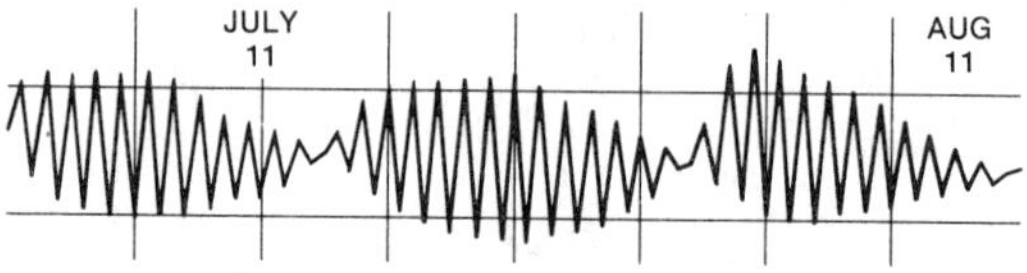

And the unusual (four times a year):

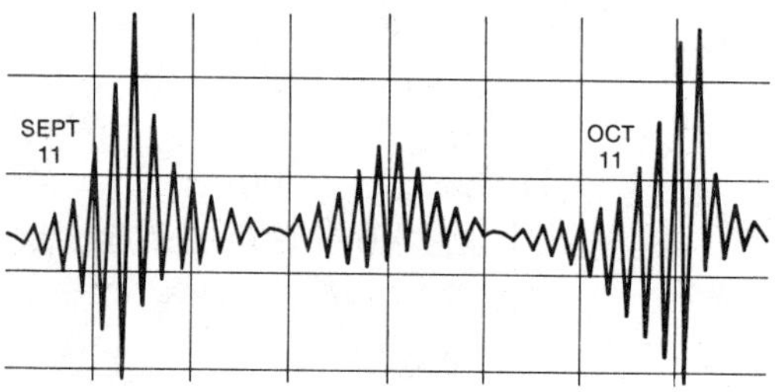

Life Is Electromagnetic. It Relates to Geophysics

Because the life sciences are rooted in "chemical" thinking, there is as yet little appreciation of the impending revolution which will reveal the fundamental electromagnetic processes of life. This will be reviewed in another appendix.

To the extent that a chemist is not inclined to an awareness that life might relate to "fields" such as gravity or magnetics, he therefore tends to discount statements that life is "tuned" to and relates to a rhythmic geophysical environment. He prefers to stick to chemical reactions. But the direct correspondence has been often shown.

Besides those mentioned in Chapter 1, there are many other experiments which show evidence of a one-to-one relationship between the biological and geophysical rhythms. For ten years, Dr. F. A. Brown (see References) kept a continuous monitoring of the respiration of potatoes, and the solar and lunar beats were clearly represented in the recordings.

A more recent article (*Science* 198:56) describes how eight monkeys were hooked up with catheters for monitoring their blood cortisol rhythms. It was found that, besides the 85–90-minute rhythm found in sleep studies, there were also peaks at 50 and 200 minutes. The authors pointed out that these could represent harmonics of the daily rhythm (45 and 180 minutes) but failed to see that it could better be understood as harmonics of the 24.8-hour lunar day.

In spite of all the evidence of life's involvement in a rhythmic environment, most researchers continue with what we consider a very limited and myopic view.

What's the Matter with Science?

Being blind to the geophysical realities may be intrinsic to the "Robot View" I have described earlier. The biologists cannot "see" the boundaries of life extending out magnetically without disturbing their mechanistic "victim" philosophy. That would bring them in contact with their full self-responsibility. We can imagine that, as children, these hard core scientists were terrorized into fearing "god," and they identify that terror with the fall into becoming conditional and non-responsible. Now they seek assurance in science of freedom from that terror. Their attitude toward circadian rhythms can therefore be just one aspect of this whole problem of how they should relate to the universe.

This blindness is reflected in a statement in the *New Columbia Encyclopedia:* "In the absence of external cues, the internal rhythms gradually drift out of phase with the environment." But hold on! If some rhythm drifts from the 24-hour day to the 24.8-hour day—is not that environmental?

I asked Dr. Aschoff, one of the leading authorities on circadian rhythms, why he was so sure that the rhythms were endogenous (internal) rather than exogenous, or geophysical, in origin. He replied that they don't, first of all, correspond one-to-one with environmental rhythms. (I would say that this base trigger can be deflected by other factors.) Secondly, because they are triggered by light, they should cease in a constant environment, but don't. My answer is that he ignores the continually oscillating gravitational and magnetic fields which permeate those so-called "constant environments."

Literally thousands of experiments have been performed in the effort to discover and/or prove the internal source of the rhythms. This consists of distorting various tissues or functions to see what effect this has on them. From single-celled animals to crabs' eyes, beetles' abdomens, rats' brains or to monkeys' adrenals, they poke, prod, and jab with surgery, drugs, starving, electric stimulation or thermal or other shocks, to see what happens. The rhythms are everywhere, manifesting in a bewildering number of ways because, instead of observing them "free running," they are usually subject to modifying conditions.

Every so often someone proposes another cellular structure, tissue, or organ as the origin of the clock. Candidates now include the pineal gland, the supraoptic chiasma, the fascia dentata of the hippocampus (of the brain), and even the protein synthesis mechanisms in eukaryotic ribosomes! But the "oscillators" are everywhere. Ignored, for the most part, is the one big oscillator up there, the geophysical generator, upheaving the whole earth rhythmically.

There Are Complicating Data

There are many facts about the circadian rhythms which are genuinely confusing to the simple one-to-one geophysical model, which, though primary, may obviously be distorted by other factors. The following are just a few examples:

Debrained silk moths emerge at random times, day or night, but when brains are implanted loosely in their abdomens, they begin to emerge rhythmically. Cyanide inhibition of enzyme activity does not stop the rhythm. Photosynthesis continues rhythmically even when algae have their nuclei removed. The eye of a crab, removed from the body, continues to manifest rhythmic neural activity, even while the rest of the body continues likewise.

A single cell may manifest four different biochemical rhythms, all circadian, all out of phase with each other. Bacteria, without mitochondria, may not be rhythmic. Phases can be shifted to any time of the day. The rhythms can vary around the geophysical rhythms without corresponding exactly. The rhythms can be modified, within limits, by temperature changes and chemicals.

If you raise fruit flies in a constant, controlled environment for as much as 25 generations in which they manifest no periodicity, and then give them one burst of light, the entire population of fruit flies will be triggered into initiating rhythmic activity. Light is a fundamental parameter throughout the whole subject that raises many questions about fundamental physics of life, light, and gravity and their interrelationships.

I am convinced that there would be little use in my wading through the enormous and contradictory test results, with all their variations in conditions, to attempt a final explanation. Much of the data (periods of 27 hours for bean leaf movements, 22 hours for algae spore release, 23.3 hours for a growth-rate rhythm in oats) must be considered suspect and in need of review. Consider, for example, that the rhythm of body temperature in the bat was found to be 22.4 hours in summer but 25 hours in winter, under the same "constant conditions."

It is evident that these controlled environments cannot take into consideration all the geophysical parameters, such as gravitational and magnetic field rhythms, seasonal variations of vertical electrical potentials, sun spots, etc.

In the final sense, Psycles does not depend on the solution to these mysteries.

References:

Aschoff, J. (ed.) *Circadian Clocks*. Amsterdam, Holland: N. Holland Publishing Co., 1965.

Brown, F. A., W. Hastings, and J. D. Palmer. *The Biological Clock: Two Views*. New York: Academic Press, 1970.

Bunning, E. *The Physiological Clock*. Rev. 2nd ed. New York: Springer-Verlag, 1967.

Gauquelin, Michel. *The Cosmic Clocks*. Chicago. Henry Regnery Co., 1967

Halberg, Franz. *Chronobiology*. Annual Review of Physiology, 31:675-725, 1969.

——— President's Letter: Chronobiology in 1977, *Chronobiologia* 4,255.1977.

Harker, J. E. *The Physiology of Diurnal Rhythms*. London: Cambridge University Press, 1964.

Luce, Gay Gaer. *Biological Rhythms in Human and Animal Physiology*. New York: Dover Publications, Inc., 1971.

Pittendrigh, C. S. *Circadian Rhythms and the Circadian Organization of Living Systems*, Symposia on Quantitative Biology, 25:159-182. New York: Long Island Biological Association, 1960.

Richter, Curt Paul. *Biological Clocks in Medicine and Psychiatry*. Springfield, Ill.: Charles C. Thomas, Publisher, 1965.

Stroebel, C. F. "Behavioral Aspects of Circadian Rhythms" in Zubin, J., and Hunt, H. F., (eds.) *Comparative Psychopathology.* New York: Grune and Stratton, 1967.

Sweeney, Beatrice M. *Rhythmic Phenomena in Plants.* Experimental Botany Series of Monographs. New York: Academic Press, 1969.

Ward, Richie R., *The Living Clocks,* New York: A Mentor Book, New American Library, 1972.

C. Psycles and Two Diary Fragments

If you don't rush right down and buy yourself a diary, you're doing yourself a great injustice. Diary-keeping is a key function of Psycles. Here we'll illustrate with two fragments of diaries how a Psycles analysis can provide revolutionary insight into subconscious realities, free you from painful bindings, and boost you toward that supreme goal of self-liberation and the perfect freedom that you long for.

Diary Fragment (A)

FRIDAY	7 A.M.	up
	8 A.M.	work
	10 A.M.	troubles pressure testing, monthly report late, boss impatient
	12 M.	lunch cafeteria, Dept. 7 gang
	aft	regular
	5 P.M.	beer with the boys on the way home
	6 P.M.	home
	7 P.M.	dinner
	8–9 P.M.	TV—Evel Knievel
	9–10 P.M.	TV—Nova on Genetics
	11 P.M.	to Jim and Jean's, booze
	1 A.M.	big argument on politics
	2 A.M.	home and bed
SATURDAY	8 A.M.	up. slight hangover
	9:30 A.M.	working on car, puttering in shop
	10:30 A.M.	(wife) annoyed, get going with wash
	11–12 A.M.	laundromat
	12 M.	lunch—san
	1:30 P.M.	sneezing, and nausea in stomach. OK after ½ hour.
	2–4 P.M.	to park with the kids, swimming
	6:30 P.M.	dinner
	7–9 P.M.	TV—?
	11–12 P.M.	Bob & June drop in, home brew beer and jiving
	11:05 P.M.	1 bottle of beer blew, all over

Sunday	1 A.M.	late-late TV, western
	2:30 A.M.	bed
	9:30 A.M.	up . . . breakfast . . . newspaper
	10:30 A.M.	sudden sharp lumbar pain, lower back, real sore. to bed, heat pads, vit. C., juices
	5 P.M.	up, better
	6 P.M.	dinner
	7–8 P.M.	TV—cops and robbers
	8–9 P.M.	TV—Cannon
	9–10 P.M.	TV—Upst.-Downstairs
	10:15 P.M.	tea and snacks—ginger bread
	11:30 P.M.	bed. back not too good

What the Diary Tells

This sequence of three days has three case histories which are uniquely revealed by Psycles. There is one "accident" and two "psychosomatics," each with its specific triggering incidents, and each revealing subconscious truths.

Looking at the events, making some connections, and seeing something of their meaning are all fundamentally liberating.

a. *The Beer Bottle Burst* Going back about 37 hours, we find a tension situation in which the boss is impatient over the late monthly report, aggravated by the fact that he's having trouble with some pressure-testing. For him to release this built-up psychological pressure, and let it loose, symbolically, he selected the right bottle from his home-brew stock, shook it just right, and struck it just right, so that it exploded, spewing beer all over the kitchen. This was a release, psychosomatically.

Recognizing the connection, he can now go back and see that the situation contained those elements, that he was bottling up his feelings, and that he needs to do something about his working situation—if it's that critical—and also something about himself and his feelings toward his boss. Does he wish the boss to "ride" him, and does his boss stand in for his "old man"? Is this something he needed from his father, subconsciously? In this case we don't say that the bursting bottle is the solution to a problem. It merely opens up part of his situation, to pinpoint where he needs to look at things and relationships.It is a precision indicator of a trouble spot that calls for attention.

In this story there is also a midpoint, Saturday morning, when his wife was putting pressure on him as well. So the bottle-burst was homeostatic for both incidents. And he must ask: why did he want her to lean on him? Is there any relationship in his mind between the boss and her?

b. *The Backache.* Sunday morning, about 10:30, he suddenly turned on a severe lower backache, a lumbar somatic. He was not doing any heavy lifting, and even if he had, it would not have been the true cause, but just an engineered "good reason." Going back 37½ hours, we find the more plausible and deeper reasons for that painful condition: Friday evening, from 8 to 9, he watched a TV program featuring Evel Knievel doing his death-defying leaps on motorcycles. During the latter part of the program there were scenes of his crushing falls in slow motion, showing how he was repeatedly injured, but never quite killed.

Was that lumbar somatic associated with some specific and terrifying childhood incident, some near-brush with death? Do we see the hooded skeleton with scythe threatening to come and grab him and take him away to some terrible fate like hell or hades? Might that pain-feeling itself be associated with the surviving of it, and mean the guarantee of life? In other words, was there some experience, long ago, in which he concluded that he would not die if that feeling were present?

This case also has an in-between event, Saturday morning, when we find him tinkering with his car as a sort of mother-womb-security structure that insures life and safety. In other words, the work on the car was not meaningless. By virtue of its timing, we see it as full of meaning, part of his deeper equilibrium efforts. It was related to the disturbances he felt Friday evening, watching the daredevil motorcyclist playing on the edge of death.

c. *The Sneezing and Nausea.* For this we only have to go back one half-cycle, 12½ hours, to the evening before. There we find him in a big argument on politics in which he felt rejected. Sneezing and nausea are typical reactions to feelings of rejection. We have hundreds of cases like this.

He didn't like to feel put down and rejected for his views, and when it happened, it was as if Jim and Jean were rejecting him personally, rather than just his views. So he took that rejection, turned it around and expelled the whole put-down. Or something like that.

To go back and see this sequence is to realize that he really felt hurt, in spite of his cover-up. And it's important to know that something was really getting to him and that he would react this way. What did he really believe about himself, and how important is it for someone else to agree with him? In the future can he argue about politics with more detachment, with less gut-caring, or was there something special about the subject that hurt? Was he identified with the target of criticism, or the candidate? If so, this is important to know, and to work on as a conditional problem.

A Reminder of Our Uniqueness

It can't be repeated too often, that each of us is a unique creature. What happened for him is not what would necessarily happen to you. Although there are some similarities, it must be emphasized that each person has his own unique background of experience which is the basis for unique responses to upsets. These are merely examples of upsets and responses for one particular person.

By keeping a diary and using Psycles, you can find out for yourself how you maintain your equilibrium, and thus act to liberate yourself from your painful past.

Diary Fragment (A)—A Psycles Analysis

FRIDAY	10 A.M.	Impatient boss troubles
	8–9 P.M.	TV: Evel Knievel—death-defying
SATURDAY	1 A.M.	Argument on politics, felt put-down, rejected
	9:30 A.M.	Puttering on car in shop
	10:30 A.M.	Wife impatient: get going on wash
	1:30 P.M.	*SNEEZING, stomach NAUSEA*
SUNDAY	11 P.M.	Beer *bottle EXPLODES*
	10:30 A.M.	sudden *lumbar PAIN,* locks in

Diary Fragment (B)

This series of events, recorded especially for our research, provides a good example of Psycles. There are two separate series, interspersed, taking place in southern California in October of 1960. The subject was an unmarried secretary working in aerospace.

FRIDAY

5:30 P.M. Going to (favorite bar) with one of the girls from work to meet some of the fellows from work for a few drinks. Also my boyfriend (BF) and a friend of his are to meet us there later. Girl I am with is to go home after they arrive, and my BF, his friend, and I are to pick up another girlfriend of mine for a double date. We are in a hurry to get there; we had to stop some place on the way, and it's a long distance to the bar. "Happy Hours" are from 4:30 to 6:30 when drinks are half-price.

6:20 P.M. Arrive at bar, each orders four drinks in advance (cheaper that way), since "Happy Hours" is almost over. Girlfriend drinks four drinks too fast, and gets sick (drunk). BF and other fellows arrive. We wait until GF sobers up a little, and then she goes home.

8:30 P.M. Pick up other girlfriend and go get something to eat. My BF acting peculiar, seems to be mad at me. Very antisocial at restaurant. (Find out later that he's mad at me because he claims I "talked too much" at restaurant.)

10:00 P.M. All four go to cocktail lounge at bowling alley. Run into married couple—old friends of mine I hadn't seen for months. Wife had been my best friend since age 14, but doesn't call me much anymore.

12:00 P.M. Girlfriend's date not feeling well (ulcers), decides he wants to go home. We take girlfriend home first. Her date wants directions on easiest way home, and BF and I get into big argument on location of a particular street. I become embarrassed at his loudness and obnoxious tone of voice in front of friends. (He's still acting bitter since time in restaurant.) Argument ceases and we depart from friends. Balance of evening vague. And troublesome.

SATURDAY

2:00 P.M. BF comes over. Argument develops again over location of that street, also location of a particular city. This leads to disagreement over toll calls between different cities in area. I tell him he is either stubborn and won't admit his error, or he's just plain stupid. I say this repeatedly, "You're either stubborn or stupid." He finally gets angry and walks out door. Not knowing if he is coming back that evening, I call girlfriend and make arrangements to go out that evening.

7:00 P.M. Getting ready to go out with girlfriend, phone rings. I figure it's BF and don't answer.

7:15 P.M. BF knocks on door. My daughter runs to door and calls, "Who's there?" I originally intended not to answer door, but daughter gives us away. I answer door and tell him a lie—I was sleeping. I tell him I'm going out with girlfriend. BF angry, says, "We're through!" and leaves in uproar.

8:30 P.M. Pick up girlfriend and go dancing. Have a very boring time.

SUNDAY

2:00 A.M. Go to drive-in for something to eat. Car next to us has four young "hoodlums" who try to start conversation. GF and I ignore them. They become insulting and try to climb in car. I become angry and order them out. Guys get out, but still making dirty remarks. Police car in drive-in. I report their activities to policemen. They search guys, and then order them off lot.

3:00 A.M. Take girlfriend home, and then go home myself. Very tired. Anxious to go to bed. Turn off light in living room, and while walking through in darkness to bedroom, I stub toe on end table and break it. Go to bed in terrible pain.

Later. 7–8 A.M. Have disturbing dream. Girlfriend and I on the way home from dancing. She starts calling me names, stupid! etc. In dream I go about ten miles too far in taking her home. She continues to yell at me, and says she'd just as soon walk home, the whole ten miles. At that time (in dream) car breaks down, just blocks from my own house. (Lug nuts shearing off left front wheel. This had actually happened a month before and BF had fixed it.) I tell her to get out of car because I can't take her home now anyway. She gets out and proceeds to walk home. I just make it to my own house.

1:00 P.M. Wake up, thinking the dream is real. Terrible pain in toe. Try to walk. Thinking of GF walking ten miles at 3 A.M., feel responsible for whole thing. Wonder if I should call her to see if she got home all right, though she is mad. Then realize it was only a dream. Remember taking her home. Decide to call her and tell her of strange dream and see if she has a car available to take me to hospital for my toe. Can't walk, so I know I can't drive. GF has no car, so I have to think of another way to get to hospital.

Decide to call BF for help. (I'm unhappy about argument, and want to make up anyway. Good excuse.) BF very distant on phone, but says he'll take me. At the hospital we make up. He invites me to sister's for dinner.

5:00 P.M. Go to BF's sister's house. Have dinner and leave. Go home.

7:30 P.M. Doing dishes. BF in living room. I break glass and cut hand badly. BF doctors hand, but says I should go to hospital for stitches. Tendon may be cut. I refuse. Once a day to the hospital is enough. I feel faint, so lie down for awhile.

MONDAY

8:00 A.M. Going to work. Toe and hand still painful. Bandages make it awkward to drive. Run into unusual traffic, late to work. Feeling nervous and upset. Hard to drive. I'm late. Hand and toe hurt. Trying to hurry. Small parking place in lot. Trying to park car clumsily. Scrape MG car parked in next space. Pull out and try new parking space.

Some Interpretative Hints

Again we have extracted the key elements and plotted them in chart form so that the various connections can be seen according to Psycles. There is value in just seeing how the events connect up, regardless of attempts to explain or interpret (in which there is much latitude). So for your own diaries, don't worry too much about the deeper meanings. The benefits are automatic.

Obviously, in this case, there is considerable role conflict between the subject and her boyfriend. The three accidents are rich in symbolism. Breaking her toe and cutting her hand may both be self-punishment for the abusive kicking and slapping she meted out to him (note the timings). And the hoodlum scene at the drive-in may indicate that, subconsciously, she felt that her boyfriend behaved like a hoodlum in entering her body early Sunday morning, when they were both hostile and inebriated.

Thus it is that people all around us (besides ourselves) are continuously maintaining their equilibrium, and balancing the books for the images we have identified with.

Diary Fragment(B)—A Psycles Analysis

FRIDAY	7:00 A.M.	?
	4–5 P.M.	Complex plans for eve. Hurry, hurry, late.
	6:20 P.M.	Late to bar, get drunk, loud.
	8:30 P.M.	BF mad at subject
SATURDAY	12–1 A.M.	Argument with BF. "Stupid!"
	2–3 A.M.	Intimate hostilities
	7–8 A.M.	? sleeping? depart?
	2–2:30 P.M.	More "stupid"arguments. BF leaves in huff.
	7:30 P.M.	BF arrives, is rejected, "We're through!" Out with girlfriend, dull.
SUNDAY	2 A.M.	Troubles with hoodlums, get police help.
	3 A.M.	*BREAKS TOE*, goes to bed in pain.
	8 A.M.	? Dream: car breakdown, calls GF stupid, walks home.
	3–4 P.M.	Calls BF for help to hospital. Make up.
	7:30 P.M.	Breaks glass, *CUTS HAND BADLY*. Faint.
MONDAY	8 A.M.	Late to work. *SCRAPES CAR* in parking lot.

D. The "Miracles" of Faith Healing

Many cases with complete medical documentation have shown that it is possible to effect almost instantaneous healing of disease. In addition to the obvious psychosomatic types, there are also such extreme organic distortions as the following:

a) bone disease which had completely paralyzed both legs, accompanied by violent convulsions, considered incurable; b) blindness with white atrophy of optic nerves, of cerebral cause, with white pearly papilla devoid of all color, considered incurable; c) advanced cancer of the breast,pronounced inoperable; d) face half eaten away by a tuberculous skin disease; e) enormous fibroid tumor of the uterus, with pressure causing chronic gastric troubles and vomiting, considered hopeless; f) tuberculosis of the spine, nephritis, and tuberculous arthritis of left shoulder with five large draining abcesses, considered incurable; g) tubercular spine, bedridden for four years, stiff as a board, considered incurable; h) normal child of five, stricken with acute meningo-encephalitis, paralyzing both arms and legs, causing frequent convulsions and epileptic fits, resulting in complete idiocy—duration wo years, considered hopeless; i) polyneuritis, involving particularly the spinal nerves and spinal cord, with legs becoming gangrenous, greatly swollen, with rest of body as thin as a skeleton, hopeless; j) tubercular peritonitis, with swollen and hard abdomen.

All of these cases, and many others, were completely and permanently cured within the space of a few hours, or a day or two, at Lourdes, France, where the official medical bureau has an enrollment of 5,000 doctors from 30 countries who assist in the methodical checking of alleged cures at this religious shrine. (See *The Miracles of Lourdes* by Ruth Cranston, 1955).

How can a terrible disease, with gross organic derangements, vanish within a few hours? Cannot these "miracles" be explained or understood?

Officially, the medical profession uses such phrases as "contrary to known medical or biological laws," "scientifically inexplicable," "outside the laws of nature," "no natural or scientific explanation," or "on the margin of the laws of nature."

As summarized by Dr. Alexis Carrel, "Patients have been cured almost instantaneously of various afflictions such as peritoneal or bone tuberculosis, abcesses, osteitis, suppurating wounds, cancers, etc. In a few seconds, or at most a few hours, wounds are cicatrized, pathological symptoms disappear, and appetite returns. The miracle is characterized by an extreme acceleration of the processes of organic repair. No scientific hypothesis up to the present accounts for the phenomena, but the only condition indispensable for its occurrence is prayer. The patient does not need himself to pray, or to have any religious faith; but someone around him must be in a state of prayer."

Now—why all this reticence to understand such phenomena?

Let's summarize, again, what we know from our revolutionary discoveries stemming especially from the application of the circadian half-wave, Psycles, to human troubles:

a) We are self-determined; we engineer our destinies.

b) We modify ourselves in accordance with our resonant identities, creating our accidents and illnesses.

c) We do what we want, and for precise conditional reasons.

d) It takes effort and energy to maintain an imperfect state.

e) We are "social" and require social permits to change ourselves.

Thus, every one of these patients was *doing* this condition to himself, for reasons, and expending much energy doing so. Each patient had exhausted his particular reasons, his particular identity and motivation, and each was approaching death. Each patient had made the decision, subconsciously, to quit it. Enough was enough. He had a new reason for living.

Lourdes was the social permit to change. At Lourdes the atmosphere is in contrast to society in general—it *allows* the "miracle."

So the miracle was that each patient, in that permissive environment, *stopped* doing what he was doing. The old reasons and motivations were dropped and subordinated to a generalized love-accept-relax-let go.

It takes constant energy and effort to de-form the body-mind. When this de-forming effort stops, the body-mind springs back automatically. The moment there is the cessation of the efforts to maintain it off-balanced, deformed, or mis-aligned, there occurs an automatic restoration of wholeness, a realignment of the energy-system, to its original configuration. The speed is fantastic. A magnetic field can be deformed by conducting elements, but returns instantly to its spherical configuration when the distorting elements are removed.

We repeat:

It takes constant energy, constant effort, to de-form the body-mind. Remove the *real* cause, the self-destructive desire and motivation, by an overwhelming change of identity—add full permission to change—and homeostasis happens.

That is the "miracle."

In fact, it might be said that the miracle is the only truly natural event that happens. All our troubles we create are un-natural, or evidences of de-forming. And when we stop our continual efforts to be like this or that, we may be automatically restored to the only completely *natural* state—the state of perfect health, un-deformed, un-modified, and un-conditional.

There are hundred of "permission systems" whereby we can manifest our unique omniscience and self-responsibility, and just do, or know, or quit something. This includes Ouija Boards, Crystal Balls, Tea Leaves, Tarot Cards, the I Ching, Hypnosis, Astrology, the Subud Latihan, the Quaker Meeting, Psychodrama, Prayer, Meditation, Laying on of Hands, Psychic Healing, and so on, and so on . . . .

Let's say I go to an astrologer or psychic. I have a request, tell me something I don't know, or, by some magical way, stop this pain. Then follow some of the permission systems, involving ceremonies and ritual procedures, cards, instruments or incantations.

And then I get what I want. Always. Did I just wish to be entertained, with the revelation of some little secret information drawn from my subconscious? Or did I wish to be disappointed once more? Or did I wish to satisfy myself that this psychic really didn't know much? Or did I turn off my pain, with a sense of gratitude to divine providence?

Whatever happens is in accordance with my purpose and need. What really happens is that we are both subconsciously omniscient. The psychic knows me inside out, at a gut level. What do I want to be revealed? The psychic will cooperate, automatically. We both agree to play a game. We are both "doing our thing," in confluence, with the psychic fulfilling his or her needs, doing the performance. There's nothing "wrong" about this. We both come together to satisfy our mutual needs. And among my needs may be a system of permission whereby I may stop some pain, or make some change in my behavior or feelings.

Although we may wish to pretend that someone Else, or some Power emanating from Above, enters and heals us, the divine miracle of it all is that we, in our divine omniscience, do our troubles for reasons, and quit them when we wish.

The "power" of the psychic healer is not in his or her hands. The energy fied emanating from skin and fingertips may affect enzyme activities—but it is minuscule compared with the total body-mind system *we* are. The power is in *our* decision to change what we're doing. We want to pretend that we're not responsible.

We want to pretend that some divine energy *out there* did it. What foolish games we play!

As Jesus said, "The Kingdom of Heaven is *within you!*" And as the Buddha said, "Nirvana is the turning-about in the deepest self!"

Miracles happen because our total body-minds are One with the Cosmos.

So—use Psycles as a precision Master-key. It can help you unlock those decisions and identities which stand between you and your ultimate Freedom!

E. Psycles and the Therapist

From religious ecstasy to well-being, to the dull, the angry, the insane, and the near-dead, people manifest in many ways. For all this spectrum of behavior there is no generally accepted set of explanations, no single system of understanding. There are many theories, many guesses, much speculation, mountains of half-truths and oceans of nonsense.

Like the Rosetta Stone

Into this scene comes Psycles, not as a singular answer, nor as the final truth, but as a unique Master-key not unlike the famous Rosetta Stone of archeology. For a long time there were several ancient languages in the Middle East that no one could decipher. Then came the discovery by Napoleon's troops of a basalt slab, inscribed by priests of Ptolemy V with the same message in three languages, including Egyptian hieroglyphics. The code was broken.

So it is with Psycles. The singular timing which exposes the specific triggers for the spectrum of behavior, from sneezes to accidents to cancer and sudden death, unlocks a thousand mysteries in psychology and the health sciences. With its one geophysical constant, that giveaway timing of the circadian rhythm, it cuts across the lines of a thousand theories to bring order out of chaos.

But as with the Rosetta Stone, Psycles is only the beginning. Although the circadian code has been broken, much remains to be deciphered. There are many applications to be explored, and mysteries to be answered.

This is not to say that we should sit back and wait for the next phases of the scientific process to take place. We don't need to let years go by while the formal research, with all the controlled experimentation, documentation, computer analysis and so forth, is undertaken.

We should put to use what is already known: with Psycles we can unlock the deepest troubles, reveal the hidden reasons for them, and gradually free ourselves from the subconscious chains with which we bind ourselves to a very limited and suffering existence. With minimum effort, Psycles can give instant self-insight to help solve personal problems and hang-ups; it can sometimes turn off serious physical symptoms in seconds or minutes; with it we can predict, to a certain extent, serious illnesses or accidents days in advance, to within minutes; it can save lives endangered by heart attacks, sudden psychosis, and other crisis situations; it can reveal the secret reasons for accidents and help to prevent future ones.

More academically, it uncovers the vast and universal ESP which underlies our relationships with other people and animals; it shows that accidents are usually on purpose and why they are done; it proves how responsible we are for our destinies and demonstrates a unitive reality which permeates all life.

The Client-Therapist Relationship

As a therapist, do you relate to clients as if they were fools that you, in your great wisdom, can help?

Or do you realize that you and your clients are both in the same conditional boat, as "blind sleepwalkers" in a sense, yet both with the same subconscious omniscience? Do you realize that they know, at a deeper level, exactly what they are doing, why they are doing it, and how it is purposive in accordance with how they identified and decided to be?

Do you assume also an automatic cooperativeness among people, a flowing-together of interests, motives, and needs, because they may have come to you knowing, subconsciously, that their problem ties in with some of yours, and that between the two of you, therefore, the adventure could be mutually rewarding?

The problem is they would like to change things, but find they are locked into a structure that has been heavily fortified with survival significance. They would like to quit the pain (whatever the troublesome behavior) but need help.

What, then, are the options?

1) The client, without help, can just quit the trouble.

2) The client can ask for help in just turning off the trouble, without going into the roots of the problem and disturbing the whole personality structure.

3) The client can say, let's dig for the roots, and adventure toward the much bigger goal.

It can be assumed that the trouble wasn't serving its purpose, or had fulfilled its purpose, or was leading to some unwanted end (such as death) which was not the primary intention.

Option (1) is rare for a number of reasons, since society prefers to see us go through certain established permission-systems. But it does happen. An example of this was mentioned in Chapter 7, the one cancer case in which a mother was degenerating rapidly after a general metastasis and approaching death. Her two daughters got into an outrageous fight over what to do with her. In the middle of that battle she suddenly erupted with a "to hell with *both* of you!" and, retching and vomiting, disgorged a horrendous putrid mass, and got well! The doctors called it a spontaneous remission. She had changed her mind.

Option (2) seeks whatever external forces or energetic stimuli that can help to oppose and turn off the aberration. This is where medicine, drugs, and a thousand therapies come in. They are not directed at the root causes, the decisions and identifications, but merely at the immediate state of activation. How they work through reciprocal activations has been described in Chapter 3. The client is saying—turn off my pain but don't touch *me*!

Option (3) is what you're looking for! They are the adventurers who are responding to a new vision of life. For them we have formulated a sample program for self-help in Chapter 16. But this does not exclude interaction with an enlightened therapist.

The Past, Present, and Future in Therapy

There are some therapists who wish to dig deeply into the past—exclusively. There are others who, with fierce opposition to remembering the past, insist on dealing exclusively with present time. And then there are those who would orient their approach entirely to future goals, or with intentions, decisions, and efforts to revise what lies ahed. The latter two types may rely heavily on positive affirmations, ignoring past roots.

But to the extent that anyone is truly bound in his beliefs concerning one over the other, just so is he making public declaration of his ignorance of certain fundamentals.

For it is absolutely true that the past, present, and future are interlocked and interrelated in all personalities and modes of behavior. And it is equally true that breaking into this conditional structure by any one of the three modes inexorably helps to loosen up the other two, as triune cornerstones of our personality structures.

No one *uses* an event of the past, with its associations and meanings, unless it is useful for the present and holds promise for the future. By exerting efforts toward a new future goal, the past and present are certainly influenced as facets of our state of being. And by efforts to modify present-time behavior, attitudes, emotions, and thoughts, so too are the past referents, and future goals likewise influenced, softened, and modified.

Therefore the therapist who insists on dealing with only the present time or the future, probably has some terrible memories which he himself wishes to avoid with a passion.

We must remember that no aberrations are *done* without relation to the distant past. The present-time sequence of events (the trigger and the mishap response) can thus be considered as precision clues to the original experiences and identities in early childhood, where logic is established.

And because the root incidents represent survival in our minds, we may not just drop the aberrant behavior! Life is tenacious. We hang on very tenaciously to something we are doing now, although it may hurt, because it is tied to that survival meaning of the past.

Contacting the immediate causal incident, or trigger, can do much to relieve acute troubles. And in some cases the relief may border on the "miraculous." Complex symptoms and pains can vanish very quickly. Remember the throat hemorrhage in Chapter 3? In that case we went right back to the original referent when she almost died of a strep throat. Her throat quit bleeding almost instantly with the connection. But in the asthmatic case in Chapter 4, the symptoms eased but did not terminate right away. Our guess is that such a problem is not solved until the essential key to it—in birth itself—is resolved. That will take more time. And therefore, there can be occasional repeats of the asthmatic "syndrome"—with less and less severity—until it is solved.

But if there are no "miraculous" healings every time, rest assured that the contact is nevertheless healing, in a quantitative fashion. Each connection adds to a general softening up of the early painful associations. Deconditioning is quantitative, step-by-step. It all adds up in the gradual liberation of the individual bound to the past.

The Basic Questions

"When did it happen? What time was it? Or when did you first feel this pain? Let's see if we can get the exact time that it started. Maybe it will refresh your memory if we scan other things that you were doing between such and such times. . . ."

Despite protests to the contrary, you can assume that memory is pretty complete and can usually be teased out of the subconscious, especially when we're dealing with a week's time at most.

Let's say, for example, that the trouble, accident, or symptoms started 7:30 Friday morning. Going back in increments of 12½ hours, "Anything happen to you Thursday evening about seven? Any troubles? Where were you and what were you doing at about that time?"

"Now, again, how about Thursday morning about 6:30 or 7:00? And then Wednesday evening about 6:00 or thereabouts. Where were you and what was happening? How were you feeling? Who was with you? Anything troubling? Now, going back a little over three days, how about Tuesday morning, early, maybe 5:00? Does this ring a bell on anything?"

In most of our descriptions of cases throughout this book we have selected the 37-hour examples as the most dramatic and typical. It should be realized, however, that the full analysis consists of a *series* of circadian half-waves, or multiples of 12½ hours. Remember, also, that it involves considerable give-or-take. (Unfortunately this is the nature of the geophysical rhythms themselves, and we can't pretend they are any more precise.) Continue all the way back to six days and three to six hours.

Sometimes, rather than go methodically step-by-step backward, you can check the 37-hour and three-day times. If they are not immediately significant, you can ask a general question, "Any sharp trouble anytime in the past week?" And often the subject will go right to the key incident.

Questioning on Basic Identities

In Chapter 15 we touched on the mix-ups in sexual identifications and some of the reasons why sons often identify with mother, and daughters with father. If we were to confront someone and ask whether they identified with the wrong sex, biologically, they would probably clam up pretty quickly, because of the social stigmas.

In the most casual manner, therefore, ask, "Between your father and mother, who do you take after?"

The spontaneous reply is often very specific. "Oh, my mother." But this may quickly be followed by a series of qualifications. So we ask about specific interests, in art, music, hobbies, intellectual pursuits, and so forth. You would be amazed at how often there stands out a clear distinction between father and mother, and the particular one with whom the person identified.

The second round of questions, not referring to the parents, relates to "What have you done to your body?" Scan the body from head to toe, and consider all the nonoptimum conditions or functions that have occurred to the body. Diseases and troubles relating to particular parts of the body, sense organs, etc.

And here, again, when asking, "Did your father or mother have such and such a trouble or problem?" you will find striking parallels between the person and the parent he or she chose to identify with. From such revelations of primary identity, and with a tabulation of physical defects, you have mapped out the essentials of that individual's whole life problem, and brought a focus for understanding many of the specific problems needing to be resolved.

Resistance Is Natural and Common

To know about Psycles does not mean that we automatically use it. Knowledge and application are two different things. After all, most of us would not go out of our way to deliberately contact painful and embarrassing things. This takes courage and the knowledge that it would be helpful in the long run. Resistance is, therefore, fairly common.

We all have a sizable store of painful memories that we don't ordinarily wish to contact, so we find reasons for avoiding them. We tend to build up "logical" reasons why we do many things, and the application of Psycles might disturb these illusions. We don't like to be confronted with evidence that we are identified with certain people or things, or that we hate or love or propitiate to a shocking degree, or that we are manifesting the insecurities and desires of an infant.

Nor do we like to find that the "abnormal" activities we do are related to questions of self-esteem, or failures in interpersonal relationships, or to feelings of loss and rejection, and so forth. We quickly find that the secret self is not the self-image that we carry around with us. Touching on these immediate and present-time problems may not seem to be extremely traumatic. But the contact may engender resistance (and avoidance) because it is only one step removed from incidents in early life containing great terror, or pain, or feelings of gross rejection and the threat of nonsurvival.

Resistance, then, really stems from these root incidents, or prime referents, out of which we made those fateful decisions, when we "died" to our free self, when we identified, and began to structure a "personality." Therefore it requires courage, candor, and a modicum of self-honesty—besides faith and trust in the goodwill of the questioner/therapist—to confront these facts.

After all, who does not have at least one painful incident in which we did something that hurt someone we loved, and for which we abhore responsibility. Or who, among us, has not suffered at least one painful "accident" or terribly embarrassing event we would firmly consider absolutely unwanted? Feeling that resistance, wishing to avoid the painful memory, we can respond in two ways: Consciously, we can refuse to divulge that incident, pretending that nothing unpleasant happened. Or, unconsciously, we can forget, occlude, and be unable to remember.

When there is occlusion, so be it. It is not ready for contact. And there is one rule for the therapist—don't push too hard. Be gentle. Be compassionate. Recognize that you're touching on something that may have been life-or-death in the distant past. Respect the need to put it off until later. It's all a gradual process. Quantitative. Little by little. That's the way it is. Other pieces may have to be loosened up in their proper order before this problem is ready for solution.

It may not seem very important right now—to slip and fall, or break something, or suddenly feel some sharp pain, or choke, or sneeze, or make some monstrous mistake. But remember! Underneath that precision event there may be deep pain. Judge from the response to the first question or two whether it is ready for resolution.

The Built-In Safety Factor

Some people have questioned the dangers of using Psycles. But it is precisely this capacity to occlude and forget what is too painful to handle right now, that insures the appropriate. Until the time is right, it's just naturally forgotten. Walled away in the "unconscious." It's a basic safety mechanism. It protects.

When Psycles Doesn't Seem to Be Working

Before you give up in disgust, run down the following checklist of reasons why you may have failed to contact significant motive behind someone's trouble:

1) You missed one or more of the multiples of 12.4 hours backward. First try 37 hours, then three days plus, and then start filling in 12.4 hours, 25 hours, 49 hours, etc. Don't be too confining with the times, saying, "round about such and such a time" "give or take an hour or so."

2) It might be a rare six-day-plus case, with an unusually long gap before the response. Most homeostatic responses are fulfilled within the three-plus-day interval. But, sometimes it stretches out, and we don't know why.

3) Twice a year (note those charts) the geophysics of tides are skewed by unusual extremes. Whether this accounts for some of the failures is open to question.

4) A true and total occlusion of the painful trigger incident may exist, i.e., it is too painful to contact at the present time and too disturbing to the overall self-image of the subject. You may detect this in the breathing, or in other signs of tension and resistance.

5) There may be confluential occlusion on your own part. Subconsciously, you may wish to avoid contact because it happens to relate to something too painful in your *own* life.

6) You missed the symbolism, the meaning, the significance of some events because it was too subtle to recognize. It took more background knowledge of the subject and his or her life before the meaningful relationship could become clear.

7) You made contact. There was the Aha! insight. But there was no sudden alleviation of symptoms. Sometimes the "turn-off" is delayed by the same Psycles timings, so that the therapeutic benefits don't occur until 12 or 37 hours later. Or, the insight was fine, but there was a lot more holding it down, and that needs more work.

F. Table of Case Histories

TRIGGER STIMULUS	*Response—37 Hours (etc.) Later*
CHAPTER 1	
The Flaming Hand	
Young father beats crying infant.	"Accidentally" burns the offending hand.
The Throat Hemorrhage	
Total rejection by angry mother.	Teen daughter starts bleeding from throat.
The Crying Jag	
Marriage prospect interested only in her body, not her mind.	Blond store detective erupts with grief.

Trigger Stimulus	*Response—37 Hours (etc.) Later*
Chapter 2—Accidents	
The Grass Whip	
Son knocks father's self-image.	Father wounds himself in ankle, in exact same spot as *his* father had big scar.
The Broken Vase	
His effort to quit smoking ridiculed by female.	He fumbles and smashes expensive vase.
The Reactor Vessel	
They argue over spaying female pup.	He breaks retort in chemical laboratory.
Two Coffee Spills	
Secretary annoyed by man in theater.	Spills coffee all over desk and papers.
Seven-year-old daughter, scared by nightmare, crawls in between parents.	Engineer spills coffee over his desk.
The Near Crash	
Speechless panic in near collision.	He topples telephone onto floor, with receiver off the hook.
The Dark Attic	
Patient resists therapist's nudge into new area of unconscious.	Accidental fall from attic trapdoor.
The Garbage Volcano	
Guru accuses commune member of leaving sloppy unclean mess.	Guru himself spills garbage over kitchen and new suede shoes.
Salesman in Hot Water	
Factory rep lets down salesman on critical order.	To raise himself out of hot tub, salesman pulls wall fixture out of wall.
Chapter 2—Car Troubles	
The Steering Trouble	
Anxiety driving home 4 P.M. with loose steering wheel, symbolizing a shaky marriage.	Waking 5 A.M., an hour too early, many mornings.
Harvey's Car	
Fears entrapment by crazy widow.	Abandons car after erratic engine trouble.

TRIGGER STIMULUS	*Response—37 Hours (etc.) Later.*
The Overheated Car	
Long distance phone call from mother with high blood pressure.	Water line pulls loose from heater, blowing geysers of steam.
A Car Lockout	
Hears sexual violence in adjoining room in hotel.	Finds himself locked out of his car.
The Lost Car Keys	
Feels shut out of wife's world.	Frantic search for car keys; locked out.
A Collision	
Expelled from board of directors.	Struck by car entering from side street.
A Near Collision	
Wife's advice belatedly accepted to avert her anger.	About to turn left, narrowly averts collision with oncoming vehicle.
Miscellaneous Car Troubles	
Wife intransigent, won't get going.	Car won't start, dead battery.
Laid off from job, felt insecure.	Given ticket for not coming to full stop.
A Highway Spin	
Acute feeling of lack of direction in life.	Car hits "black ice," hits guard rails, spins around in middle of freeway.
President Ford's Collision	
Publicity about son smoking pot.	Presidential limousine struck broadside by car full of teenagers, October 1975.
CHAPTER 2—SEXUAL PRIVACIES	
The Near Burglary	
Resists amorous lover, retains virginity.	Interrupts burglar about to steal her valuables.
The Honeymooners	
Difficult time for young married couple's first night.	Catches fly on desk, rips pants wide open.
The Lathe Bolt	
Angry re sexual advice from future father-in-law.	Machinist snaps gripper bolt, crashing large tube to floor.

Trigger Stimulus	*Response—37 Hours (etc.) Later*
The Hot Pot	
Hot date canceled with femme fatale.	Panic at overheating of titanium melt.
Boswell in London	
Lady Northumberland refuses to help.	Haggles over price, then rejects monstrous whore in pub.
The Sign post	
Ignores signs of impending end of love affair.	Runs into signpost, cuts forehead.
The Thumb Gouge	
She says I'm busy, don't want to see you, shuts door in his face.	He gouges his thumb with sharp claw-hammer while instructing students in carpentry.
The Hot Hand	
Widow holding hands with son departing for military service.	She burns same hand on hot furnace rack.
The Open Camper	
Artist indicates sexual availability to client, ignored.	Tailgate left open, paintings exposed, overnight, but not disturbed.
The Bike Detour	
She switches to date with lesbian, unsure of sexual role.	While riding bike, car makes right turn diverting her course into park.

Chapter 4

The Asthmatic Cyclist	
He severs engagement with fiancée.	Asthmatic choking, strangling, adrenalin.
The Communal Asthmatic	
Parents' crisis confrontation, to quit commune.	Severe asthmatic attack, after several intrauterine and birth symbolic events.
One Sneeze	
Hitchhiker felt rejected.	One sneeze.
Several Vomits	
Parental conflicts/rejection.	Four-year-old boy vomits in middle of night.
Artist ridiculed by fellow-artist.	She says, stop the car, gets out, vomits.
Father rejected, ousted from band.	Infant suffers nonstop vomiting.

TRIGGER STIMULUS	*Response—37 Hours (etc.) Later*
Coughing Fits	
Accused of being stingy in public	Violent, continuous, hacking cough.
Serious run-in with the boss.	Coughing fit.
Wife criticizes TV viewing habits.	Sharp coughing bout.
Diarrhea	
Rejected by girlfriend at party.	Sudden total diarrhea; finis.
Colds	
Watched goldfish flopping out of water, in experimental trauma.	Initial "cold" symptoms turn on. Birth trauma restimulated.
The Squelched Teacher	
Teacher forbidden to speak at meeting. Sat fuming.	Sudden throat congestion.
The Tardy Beauty Queen	
Suppressed anger with mother's slowness to help in panic.	Burst out with hives all over her face.
The Itchy Hindu	
Frantic anger with guru's avoidance.	Furious itching all over body.
The Engineer on Vacation	
Taunted by neighbor, suppressed anger.	Eruption of hives all over body.
Conjunctivitis from Bad Scenes	
Viewing dead body from auto wreck.	Eyes suddenly itchy and inflamed.
Earaches from Bad Phone Calls	
Three phone calls violate power role.	Burst ear drum? from swimming pool dive. Intense pain, right ear only.
Earache from the Wrong Call	
Anxiously awaited call, turns out to be commercial, unwanted.	Painful earache in right ear, 2:40 A.M.
The Fluttering Heart	
Sharp quarrel with husband.	Fluttering heart, lump in throat, apnea.
The Irregular Hearts	
Infant son threatens costly tools.	Trembling, jittery, weak, faint, irregular heart.

TRIGGER STIMULUS	*Response—37 Hours (etc.) Later*
Viewing death-defying feats of Evil Kneivel	Heart jumping, beating irregularly.
The Soviet Commissar's Heart	
Disciplined drunken sailors.	Irregular heart.
The Bursting Heart	
Fifty-five-year-old father falls hopelessly in love. Proposal rejected.	Three-inch ballooning of left ventricle.
The Hepatic Surfer	
Catastrophic failure in life goals.	Belly-up syndrome with hepatitis.
Hepatitis Among the Devotees	
Feelings of failure to attain.	Hepatitis.
Headaches	
Saw woman "who broke up marriage."	Headache, right across the eyebrows.
News: parents coming to stay.	Same.
Disturbed by police search.	Powerful headache plus "upset" feeling.
Heated argument with grown son over treatment of family.	Headache.
Ousted from business project.	Headache, neck, and shoulder pains.
Migraines	
Resented babysitting friend's kid.	Migraine plus nausea, vomiting, pains.
Left at home "stuck" with children.	Migraine.
Socked in the eye by a 14-year-old girl who didn't like his attentions.	Migraine.
Conflict with mother over body odor.	Migraine and scotoma.
The Toe Ache	
Impulse to kick husband out of bed.	Sudden toe ache, right foot.
The Sizzling Foot	
Job crisis involving Dictaphone foot pedal.	Neuritis, in right foot arch, tingling pain.

TRIGGER STIMULUS	*Response—37 Hours (etc.) Later*
The Muscle Spasms	
Fifteen-year-old daughter leaves home.	Painful spasms, like "labor pains."
Some Leg and Ankle Pains	
Intense disgust with work situation.	Sudden pain in calf, helped by walking.
The Would-be Olympian	
Deep indecision about a trip.	Sharp pain in both calves, like efforts to get moving!
CHAPTER 5	
The Hypoglycemic Engineer	
Feeling of survival threat triggered by paranoid TV drama.	Acute hypoglycemia.
The Professor Who Quit It	
Series of job-security threats.	Acute episodes of hypoglycemia.
The Hypoglycemic Divorcé	
Fear of ex-husband's violence.	Acute hypoglycemia.
CHAPTER 6	
The First Epileptic Case	
TV drama with auto-wreck reminder.	Epileptic seizure.
The Seattle Schoolboy	
Terrorized by bullies in school.	Seizures.
Bullied by his older brother.	Seizures.
The Incestuous Brother	
Visit from sister.	First seizure in four years since leaving home and living 1,000 miles away.
A Seizure in Mexico	
Sight of slow-dying horse, a reminder of husband's violence.	Seizure "blackout."
The Story of "Checkers"	
Withdrawn from school and bullies.	No seizures.
CHAPTER 7	
The Amazing Warts	
Doctor *buys* it for a penny.	Warts start disappearing.

Trigger Stimulus	*Response—37 Hours (etc.) Later*
Senator Taft's Cancer	
Three-time loser for presidency.	Self-destruct with cancer.
Ramon and His Sexual ID Problem	
Separation from wife, sexual problems.	Self-destruct with cancer.
A Spontaneous Remission	
Two daughters battle over what to do with mother.	Retching, disgorging, and termination of cancer.
How Connie Was Mis-Named	
Named after father, sexual ID problems with first pregnancy.	Suicidal self-destruct with cancer of all female organs.
Exploring Bob's Cancer in India	
Fateful decision on Sexual ID.	Start of lymphatic cancer.
Chapter 8	
My Crazy Mother	
Visited husband's grave for first time.	Psychotic break, solution to guilt feelings.
The Schizophrenic Artist	
Criticized by art teacher.	Heard voice, "Get a knife and cut your throat."
The Disoriented Carpenter	
Saw movie drama of psychotic "Jesus."	Sudden disorientation, dizzy, confused.
The Safety Inspector	
Phone news: his admired model given shock therapy.	Jittery, "about to lose my mind" feeling.
A Sudden Depression	
Tour of beautiful condominiums 55 years old, washed-up, failure.	Feeling of intense depression.
A Suicidal Impulse	
Divorcing Catholic mother returns home, finds son abducted.	Breaks down crying, threatening suicide.
On Losing Things	
Loss of uniquely loving relationship.	Loses wallet, valuable papers.

TRIGGER STIMULUS	*Response—37 Hours (etc.) Later*
The Missed Plane	
Legal exam interrupted by obnoxious female examiner.	Misses plane, theatens suit against "big mother."
The Rambling Speech	
Hears of tragic accidental death touching personally.	In public speech departs from text, to ramble on subject of death.
CHAPTER 9	
The Grandmother	
Heart-to-heart talk about death.	Peaceful exit with smile on her face.
Michael Rockefeller	
News of parents' divorce.	Death in crocodile-infested waters in New Guinea.
Max Baer	
TV special *The Killers.*	Heart attack and death.
Clark Gable	
Final filming of *The Misfits.*	Heart attack.
John F. Kennedy	
Deep identity. Gettysburg anniversary.	Shot in Dallas.
Touching a Few Other Deaths	
Joyce Funk's ID with brother.	Killed by "runaway" car.
Mike Sanger's best friend killed.	Industrial "accidental" death.
Mrs. Dana attended funeral of friend.	Suicide in Florida.
Frank Booth on way to father's funeral.	Shot by police officer.
CHAPTER 10	
Robbed by a Guru	
Psychologist's propitiations denied by guru.	Robbed of uniquely significant items.
The Little Old Lady Mugged	
Mother denied her usual birthday party.	Mugged on way back from grocery.
The Secretary Who Was Attacked	
Sexually frustrated by sleeping husband.	Gentlemanly client chases her around desk, trying to hug and kiss her.

Trigger Stimulus	*Response—37 Hours (etc.) Later*
The Rape of a Dancing Instructor	
Husband-wife marital battle.	Raped by stranger in strange part of city.
Traffic Ticket	
Laid off job, feeling insecure.	Traffic ticket for incomplete stop.
Police Brutality	
Given tongue-lashing by tough father.	Teenager sasses cops, is manhandled and thrown in jail.
Unjustly accused by divorcing wife.	Arrested on wife's complaint, provokes violence, booked.
Chapter 11	
The Uptown Nightmare	
Informed of business catastrophe.	Horrible nightmare 2 A.M.
Three Sleepless Nights	
Ski club fiasco, lonely, isolated.	Three sleepless nights.
Dream of Silver Coffins	
Dying her hair blond, feeling guilty re mother-ID.	Highly symbolic dream "mother would turn over in her grave if . . ."
The Hot Foot Dream	
Flat tires and discovery that new wheels had been put on wrong.	Dream of Fu Manchu giving her hot foot.
Seattle Triggers—New York Dreams	
TV scenes of Sahara famine.	Dream of bitch with pups and dry tits.
The Wet Dream	
Lunch date with exciting female in "Garden of Gods" in Colorado Springs.	Dream of Frenchman ejaculating, and waking with nocturnal emission.
Predictive Dreams	
Dream that package would arrive 37 hours later.	Letter arrives saying package sent.
Dream of parking lot accident.	It happens 12 hours later.
Chapter 12	
The Suicide Attempt	
Mother feels depression, anguish, and again 37 hours later.	Daughter has crisis problem, and attempts suicide 37 hours later.

Trigger Stimulus	*Response—37 Hours (etc.) Later*
Infant Tantrum	
Mother's outburst of paranoia and self-pity.	Two-year-old son erupts with tantrum.
The Bee Sting	
Couple clash, stay home New Year's.	He's stung by bee on right ring-finger.
The Suicidal Bird	
Suicidal feeling and mystique of death.	Bird dive-bombs right into radiator.
Chapter 13	
The Suicidal Fisherman	
After two previous "almosts" because marriage breaking up.	Death in plane crash, Jamaica Bay, New York.
The Research Lab Explosion	
Hell-fire sermons in church.	Hell-fire in laboratory explosion.
The Highway 101 Sniper	
Aging bachelor announces engagement.	Shot to death by highway sniper.
Chapter 15	
The Dorm Fire	
Attended androgyny seminar on bisexuality.	Incipient lesbian accidentally starts fire in dorm room at convention.

G. Disease and Healing

The 3-D Network of Neuro-Anatomy

From five large volumes of neuro-anatomy, I extracted all the circuits of the human body. Then, step-by-step, I combined them all into one picture (Figure 14). I did this because, intuitively, I felt that we are not as complicated as we like to pretend. I wanted to assure myself that the body-mind system is, in fact, much simpler than generally conceived, more integrated, more holistic, and unified. The final picture turned out to be something like a three-dimensional crystal lattice or network.

All living beings start out as a single cell. After a series of cell divisions the cell evolves into a little ball with three germ layers. From this it goes on to become all the vast variety of life forms. From an orderly series of cell divisions, and with all the symmetries we find, we should expect to end up with some structure that is comprehensible.

And so it was a pleasant surprise to discover that we are only a few cells deep—about six cells from any external surface to center. Figure 13 shows just a portion of this network, from a muscle, to adrenals, to stomach. These are not fictitious connections, but what actually exists in neuro-anatomy.

This is important. Because, when we are talking about causal chains, we are referring to cell-to-cell junctions and interconnections. In our efforts to understand behavior, it would be unreasonable to try to relate to this concept if it were too complex. If we have one bias, it is in favor of simplicity: we're not really all that complicated if we quit fooling ourselves. And this simplicity must be reflected in our structure.

It now becomes much more manageable for us to describe the mechanics of behavior. We can describe our behavior as responses (in the maintenance of a self-image equilibrium) that take place through a chain of cells, in a three-dimensional network, to bring

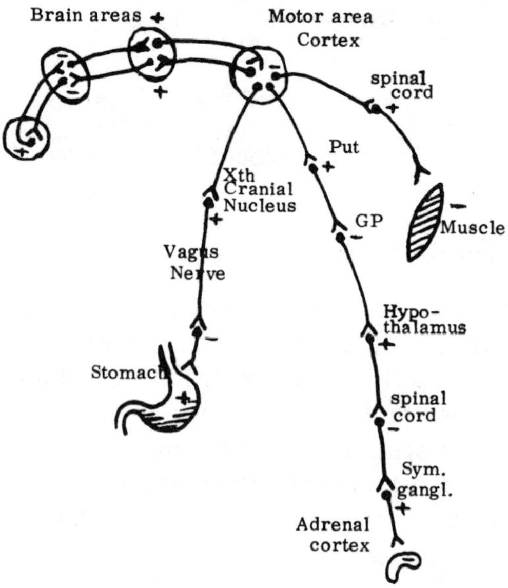

Fig. 13

about the appropriate activations and modifications of tissue cells. Recognizing the symmetries and the fact that there are relatively few cells from outside to center, we can see that we are fairly simple in construction, and easily analyzed from a structural standpoint.

This chain of reactions, however, is not like a chain of dominoes. It is not a domino theory. (That would imply that you may start something without knowing where it's going to end up.) No, in the human cellular chain of events, you are in command

Neuro-anatomical Chart

Index

AD—AnteroDorsal nucleus of Thalamus
Adr—Adrenal Medulla
Ant—Anterior Nuclei of Thalamus
AV—Antero-Ventral Nucleus of Thalamus
Caud—Caudate Neucleus-Corpus Striatum
Cerblm—Cerebellum
Cyng. Gyr.—Cingulate Gyrus
CM—Centrum Medianum of Thalamus
Coch—Cochlear Nuclei (Dors. & Vent) Medulla
Cu—Nucleus Cuneatus. Cerebellum
Deit—Deiter's Nucleus. Medulla
(Lateral Vestibular Nucleus)
Dent—Dentate Nucleus. Cerebellum
DL—DorsoLateral Nucleus. Thalamus
DM—DorsoMedial Nucleus. Thalamus
GP—Globus Pallidus. Corpus Striatum
Gr—Nucleus Gracilis. Medulla
Hb—Habenular Nucleus. PaleoThalamus
H. Thal—HypoThalamus
I. Col—Inferior Colliculus. Midbrain
(Corpora Quadrigemina)
IP-Interpeduncular Ganglion. HypoThalamus
LGB—Lateral Geniculate Bodies
LV—LateroVentral Nucleus. Thalamus
Mam—Mammillary Bodies. HypoThalamus
MC—Mesencephalic Nucleus. Medulla
MGB—Medial Geniculate Bodies
Nig—Substantia Nigra. Sub-Thalamus
PF—ParaFascicular Nucleus. Thalamus
Pit—Pituitary Body. HypoThalamus
PL—PosteroLateral Nucleus. Thalamus
Pul—Pulvinar. Thalamus
Put—Putamen. Corpus Striatum
PV—PosteroVentral Nucleus. Thalamus
PV, Sop—PariVentricular Nucleus. HypoThal.
Red—Red Nucleus. Sub-Thalamus
Resp—Respiratory Center. Medulla
Ret—Reticular Substance. Sub-Thalamus
(Corpora Quadrigemina)
S. Col—Superior Colliculus. Mid Brain
Somes—Somesthetic Area of Cerebral Cortex
Sop—SupraOptic Nucleus. HypoThalamus
SubTh—Sub-Thalamic Nucleus
Tec—Tectum. Roof of MidBrain
Teg—Tegmentum. Mid Brain. Ventral
Vestib—Vestibular Nucleus. Medulla
ZI—Zona Incerta. Sub-Thalamus

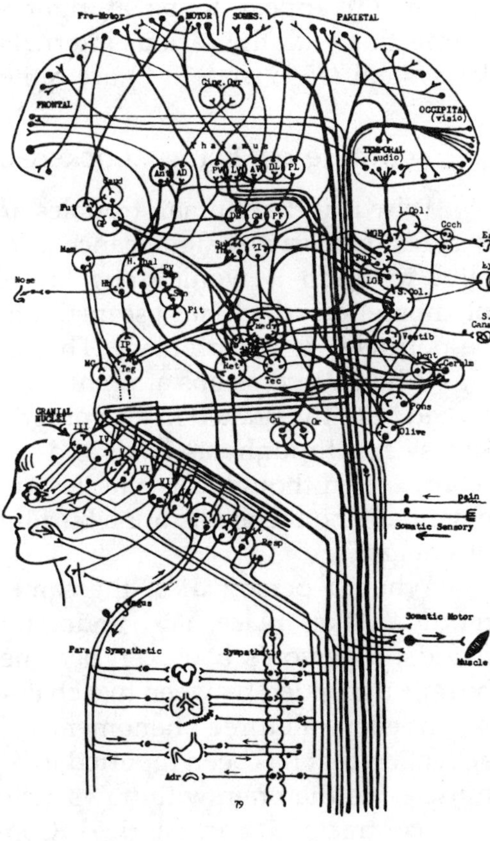

Fig. 14

both ways, throughout. This is illustrated by the plus and minus signs, for opposing states of activation.

Our Fundamental Symmetries which Underlie Reciprocal Functions

The internal organs and their nerve connections look very jumbled and complicated, but also show symmetry when cut open, spread flat, with the kinks straightened out. Figures 15 E and 15 F were derived from Figures 15 A, 15 B, 15 C, and 15 D. 15 A is the cross-section through the middle of 15 C. Figure 15 B shows the brain and central nervous system, labeled as (1) throughout. The parasympathetic nervous system (2) and the sympathetic nervous system (3) appear to be at right angles to each other, both connecting with the various internal organs (4). The blood-vascular system (15 D and 15 F) are quite obvious.

Footnote on Being "Locked-In": Tetany

Why is it that human troubles are often so difficult to undo? Besides the survival significance in the decision to *do* the trouble (and to undo it would *seem* like death), there is a physical phenomenon which has escaped attention but which I believe to be pertinent to the problem. That is the function of tetany.

When a bird sleeps at night with one leg up, the other leg is locked onto the branch in an iron grip that requires no effort. The bird sleeps through the winds and storms with one leg in a state of tetany. Even though the muscles are rigidly contracted, there is essentially no metabolism, no burning of glucose, no expenditure of energy.

When a person dies, the same phenomenon called "rigor mortis," takes place, and under hypnosis, with suitable commands, a person's body can become rigid, as stiff as a wooden beam, suspended between two chairs, impervious to heavy loads.

I believe all three phenomena are a function of the reciprocal activation that has been reported to be the case between nerve and muscle. As the brain withdraws activation, the muscle automatically contracts, to become rigid. Conversely, it takes brain-nerve *input* to relax the muscle.

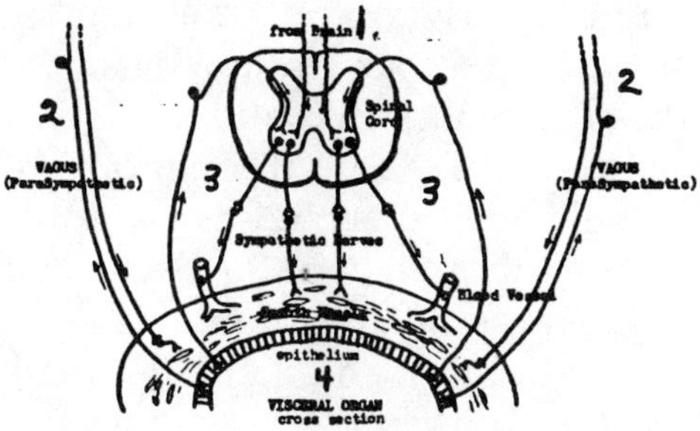

Fig. 15–A

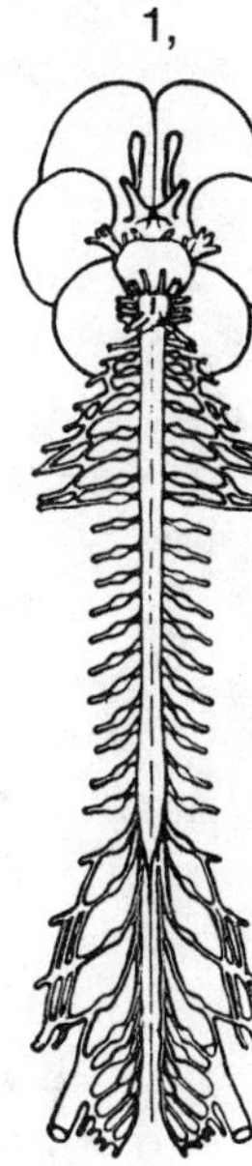

Fig. 15–B

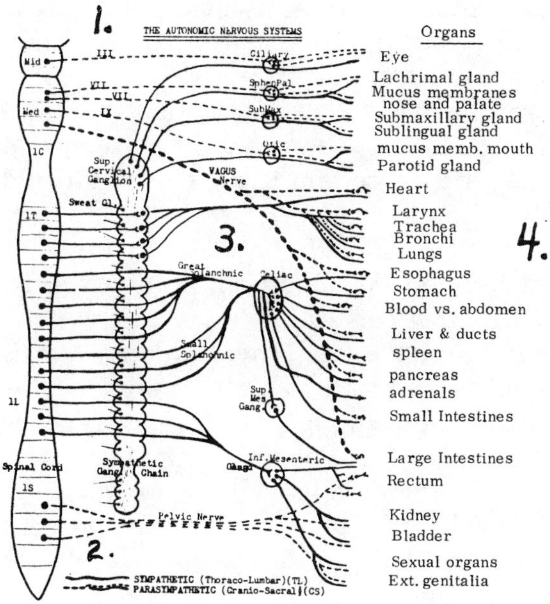

Fig. 15–C

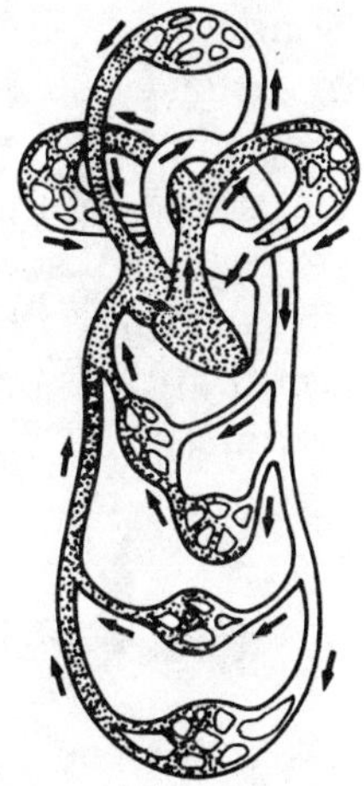

Fig. 15–D

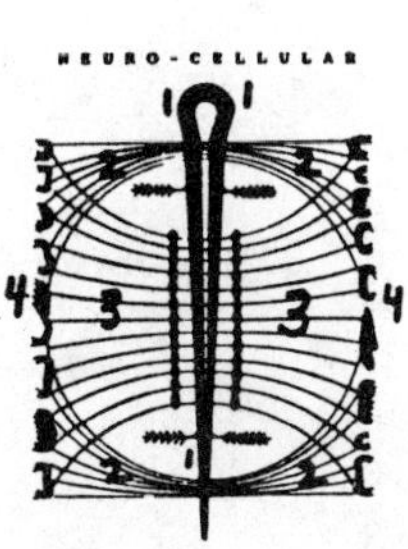

Fig. 15–E

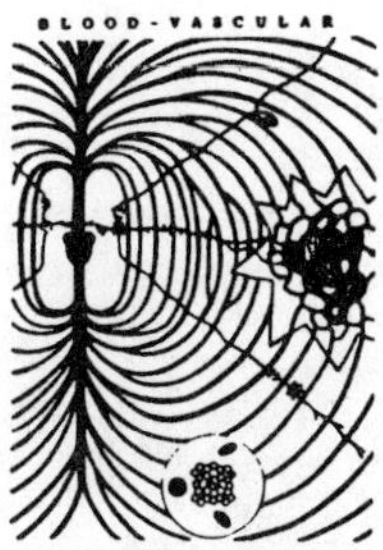

Fig. 15–F

Experimentally, a decerebrate cat (with the cerebrum removed) becomes rigid with spastic muscular contractions. But stimulation of some of the cut paths causes relaxation.

And so with the psychosomatic modifications throughout our bodies—all our "diseases": it would *seem* that they are autonomous, released from brain control. As though once the decision is made (with survival significance) to do some distortion, it is set so and so, and the brain ("I," "me") withdraws, and that little circuit spins by itself, with its own energy, unconscious and untouchable by the conscious "I."

In support of this concept there is the discovery of bacterial flagella which spin like little electric motors, in total contradiction to normal "chemical" explanations for life processes. This is part of a whole movement toward a more "electromagnetic" understanding of life. From these new premises it might be seen that "decisions" may close the loop on certain circuits, which could very well spin essentially without further "effort."

It should be noted, however, that with the snap of a finger, or a one-word command, the hypnotized subject can suddenly relax, with a complete reversal of the "distortion," i.e., the tetany. Similarly, as a biologist in India, I have seen other sudden reversals of diseased states (described as "miraculous cures") performed by the famous (avatar) Sai Baba. One young boy of eleven was in the last stages of muscular dystrophy. I had observed him closely for several weeks, draped over his father's lap, totally gelatinous, with flies occasionally walking over his eyeballs. And yet with one overt command from Sai Baba, the lad sprang to his feet to be permanently cured in one second!

Now *that's* homeostasis!

Reciprocal Functions

Throughout living bodies there can be found *pairs* of nerves, muscles, organs and systems—which are *opposed* to each other, in balance, mutually antagonistic, or reciprocal in their action.

The stimulation (whatever that means) of one means the automatic inhibition (whatever that means) of the other.

Give a *plus* sign to one, and the other is *minus*. And vice versa.

EXAMPLES OF RECIPROCAL FUNCTIONS

1. Skeletal muscle pairs—flexors and extensors—when one contracts, the other automatically relaxes.
2. Eye-muscle pairs: with automatic synchronization.
3. Submaxillary gland activation: two simultaneous reflexes:
   a) relaxation of muscular coat of arteries
   b) stimulation of secretory activity of gland cells.
4. Lung distension in breathing: exciting afferent nerves of pulmonary vagus: inhibits inspiration and excites expiration.
5. Breathing: stimulation of central ends of superior laryngeal nerve, the second branch of Vth, and the Splanchnics—causes inhibition of the inspiratory, and contraction of the expiratory muscles.
6. Pain phenomena:
   Pin jab in foot inhibits knee reflex
   Pin jabs in frontal cortex, inhibits rest of cortex—no motor reflexes
   Sham rage in decerebrate cat: No cortex, emotions uninhibited.
   Hypersensitivity in spinal cat: Higher inhibiting centers removed.
7. Autonomic Nervous System: Two reciprocal systems—
   a) Sympathetic (Thoracico-Lumbar)
   b) Parasympathetic (Cranio-Sacral)
8. Cortical-SubCortical functions: with emotional stress—
   Cortical inhibition, and thalamic stimulation.
9. Calcium and Potassium dualism: opposite effects in:
   nerve excitability, heart contractions, muscles, pH, sleep
10. Hormonal dualisms, such as adrenalin/acetyl choline. etc.
11. Sensory-Motor dualisms: afferent and efferent nerves.
    Motor and Sensory areas of Cortex
12. Visceral and Somatic dualisms:
    See details on embryology of spinal cross-section.
13. Color Vision dualisms: green/blue versus orange/red
    Correlates with: Ca/K ions; anode-cathode stim.;
    Sympathetic/Parasympathetic stim.
14. Motor Nerve and Muscle: Passage of motor impulse over a nerve does NOT, strictly speaking, stimulate a muscle to contract, but, as a result of the removal of the inhibiting effect exerted by the polarized end-plate of the motor nerve axon, the muscle automatically contracts. (*Science* 109, 289.)

Reciprocal Functions in the Autonomic Nervous Systems

Sympathetic Nerves	Parasympathetic nerves
(Thoracico-Lumbar)	(Cranio-Sacral. Vagus)
Every sympathetic ending is a miniature adrenal medulla	Liberates acetyl-choline at nerve endings.
Endings are intra-cellular	Fibers of separate origin and route independent of blood vessels.
Fibers run along blood vessels from Esophagus to Rectum	
Stomach	
Relaxes: Inhibits digestive functions. Stops peristalsis. Relaxes muscles of gut	Contracts: Stimulates digestive activity. Augments peristalsis and muscular contractions.
Heart	
Contracts: Stimulates heart activity. Produces faster and stronger beat	Relaxes: Inhibits and slows heart. Endings not in true cardiac muscle but in Aortic arch. Reflex inhibition also by PAIN stimulus anywhere.
Lungs	
Relaxes: Stimulates respiration by relaxing bronchial muscles	Contracts: Inhibits respiration by constricting bronchial muscles. Violent constrictions in asthma
Hair and Sweat Glands	
Contracts: Stimulates secretion via acetyl-choline. Erects hair muscles	No connections?
Nasal	
Contracts: nasal vessels	Relaxes: nasal vessels
Eyes	
Relaxes: Dilates pupils	Contracts: Constricts pupils
Kidneys	
Contracts: Constricts vessels inhibiting action	Relaxes: Dilates vessels, thus stimulating excretory action

BLOOD VESSELS

Contracts	Relaxes
CONTRACTS: Constricts muscles of vessel walls. Maintains tonus When cut, vessels dilate, but regain tonus later.	RELAXES: Supplies only special regions: erectile tissues, facial and glossopharyngial nerves to salivary glands, tongue, tonsils and pharynx.

SPLEEN

Contracts	Relaxes
CONTRACTS: Constricts muscles of capsules, causing exodus of erythrocytes	RELAXES: Dilates organ? also constricts muscles of veins?

GENERATIVE ORGANS

Contracts	Relaxes
CONTRACTS: Constricts muscles of blood vessels	RELAXES: Dilates blood vessels

UTERUS

Contracts	Relaxes
CONTRACTS: uterine muscles	RELAXES: uterine muscles

LIVER

Contracts	Relaxes
Inhibits? How?	Accelerates storage of glycogen Stimulates bile secretion Stimulates pancreas to produce insulin. Blood sugar drops

The Embryology of the Spinal Cord

In the early development of the spinal cord there is formed what is known as a "gastrula," which is a nearly closed, two-layer cup of cells. On top of this spherical cup is formed a "neural plate," whose two sides fold upward to form a hollow tube—the spinal cord. This extends lengthwise and one end becomes the brain.

Here it is in cross-section

Fig. 16

Now let's look at a cross-section of just the top neural tube, identifying the functions of the nerves in each of eight areas—where they connect, and what they do. (Figure 17). There are two sets of opposed, reciprocal functions: Sensory/Motor and Visceral/Somatic.

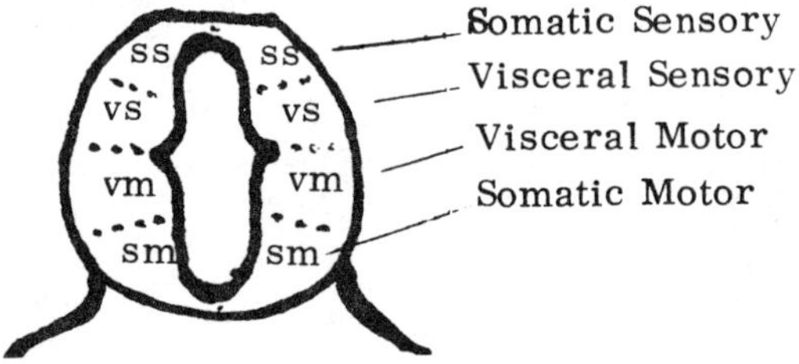

Fig. 17

Now let's spread this tube flat, and see how all these functions line up.

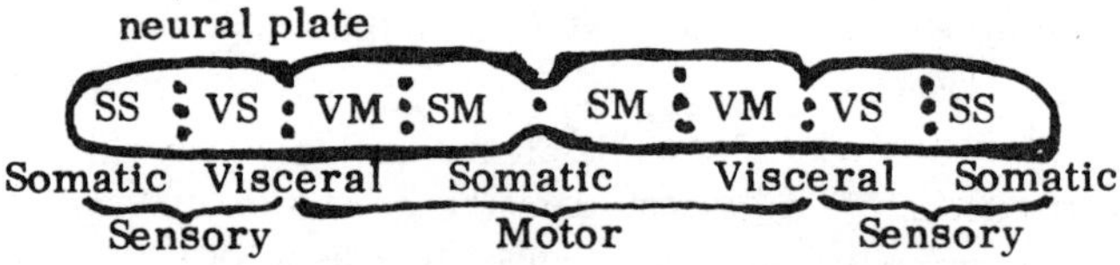

Fig. 18

Then backtrack to the original divisions:

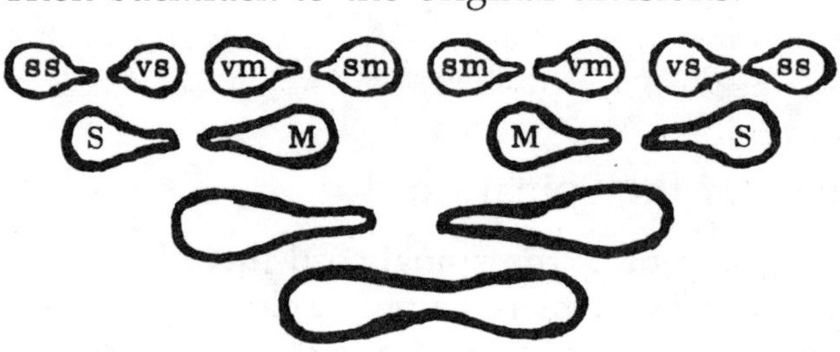

Fig. 19

This can provide remarkable insight for biologists!

What it shows is that when cells of the spinal cord divide (at least at these early stages) the daughter-cells are mirror-images *and* polar reverses *and* reciprocal in their functions. This is not generally recognized or accepted in biology, but can be considered a keystone in our simplified analysis of disease and healing. It has another great importance in that it points to the micro-physics rather than the chemistry of life. You see, such phenomena as mirror-images, polar-reverses and opposed (reciprocal) functions do not just naturally emerge from reactions in a chemical soup. We will discuss this in greater detail in the last appendix on the Secret of Life.

H. A Revolution in the Secret of Life

Elsewhere in this book there have been hints that another revolution is brewing down at the molecular level of the life sciences. There is good reason for this.

We have said that Psycles *shows* a vast subconscious awareness and communication (ESP) and a species of omniscience that permeates what we call a Unitive Reality. We have said that all this comes right out of our case histories, again and again, as we see people coming together cooperatively on the basis of such ESP.

The BioSciences do not support such phenomena. According to molecular biology, biochemistry, physiology, medicine, and psychology, such phenomena cannot exist. There must be "physical" forces in contact between a cause and an effect. Such "field" functions have no place in the chemical model of life which they all hold.

If the BioSciences are right, then Psycles is nonsense, and none of these phenomena actually exist. They are just illusions based on false assumptions.

But if Psycles is essentially correct, then the BioSciences are in need of a revolutionary upheaval so they can begin to understand what's really happening all around us.

We will, in this brief Appendix, see where they stand, what their basic beliefs about life are, the many contradictions to their data and theories, what's wrong and missing with their views, and finally—believe it or not—how life is *really* happening!

Where Are the BioSciences Right Now?

Publicly, it might seem that the life sciences are successfully solving mystery after mystery and problem after problem, secure in their knowledge of the basic processes of life, and even able to tinker with the genetic substance and create new forms of life.

Actually, they're in a terrible mess.

Don't be misled by the fact that researchers are able to do many fascinating things with living cells, to cut and splice various lengths of DNA, slip them into different cells and bring about new life forms and processes. These accomplishments may seem to infer great wisdom about life. But actually, the methods were derived hit-or-miss, by trial-and-error, even while the final knowledge of the ultrastructures and processes are quite unknown.

The one central theory in the life sciences, used to "explain" the ultimate secret of life, to describe chemical processes in living cells and to guide the billion dollars of research in biochemistry, is shot full of holes. Contradicted by a whole series of confirmed facts, it stands patently false and physically impossible and yet, like the dead hero of *El Cid,* clad in shining armor, propped up on his white horse and holding his banner high, continues to lead the troops into battle.

And the troops don't even know that their leader is dead!

The story of this incredible situation, in which some 100,000 researchers in biology and biochemistry forge ahead, chanting the same catechism, in spite of one contradiction after another, is certain to be considered one of the strangest phenomena in scientific history.

And every time they announce another discovery, in spite of untenable theories, the troops make obeisance to the central dogma, remaining conveniently blind for the most part to the frightening truth that they are still fundamentally ignorant of how things are really happening in the living cell. But cracks appear.

The "Crisis" in Biochemistry

In February of 1973, an important scientific convention brought together some 60 worldwide leaders in biology, chemistry, and physics to discuss—could you believe it—"The Crisis in Bioenergetics"?

Conference organizer David E. Green, Director of the Institute for Enzyme Research of the University of Wisconsin, pointed out that there were no acceptable theories to cover the five major mysteries of life. They are: how muscles contract, how nerves conduct, how enzymes work (enzyme catalysis), how energy is captured and utilized (energy transduction), and how electrons move (charge transfer). These questions are what life is all about.

Despite decades of research by a hundred thousand scientists with the most sophisticated instruments, all these questions still remained mysteries. As stated in *Science* 181:583, "The group were convinced that unless new dimensions were added to the conceptual framework, progress would grind to a halt." Those are strong words for scientists! The experts were facing up to the crisis of their ignorance.

Following another conference of over one hundred participants in Italy in September, 1975, Dr. Green again reported the situation as "distressing" (*Science* 193:224).

Now, in 1981, over five years later, the various BioSciences, including biochemistry and molecular biology, are proceding on course, as yet untroubled by any radically new ideas, and seemingly oblivious of the many serious contradictions which show up in experimental data to challenge their central beliefs about how life happens.

The Structure of Life

To understand how something works, you must first take it apart, piece by piece, to see its structure. After seeing something of its structure, then you may attempt to explain how it works, or its function.

Now in the case of life, microscopes weren't very good 200 years ago. Cells looked like little (membrane) bags, containing a blob (nucleus) floating in a thick soup (cell sap, or protoplasm), in which there were also some small bodies called mitochondria, golgi apparatus, and plastids. The main science was chemistry (physics was much more primitive), and it was naturally concluded that life was to be understood as complex reactions in a chemical soup. Life was chemistry.

As the years went by, and microscopes showed finer and finer detail, it finally turned out that there was almost no soup there at all, but a fine criss-cross of fibrils and microtubules, laminations, coiled coils, all interconnected most intricately. Protoplasm became obsolete. But the thinking remained chemical, with various atoms and molecules in a soup reacting with each other, somehow rightly and statistically, to bring about all the life functions.

If we could take a cell and magnify all its insides to see its ultimate structure, it would be trillions of atoms (of about a dozen kinds, or elements) stuck together in millions of molecules, forming sheets, chains, pleats, coiled coils, and tubules, all organized into beautifully intricate designs.

Before one cell can divide into two identical cells, in the whole process of growing and reproducing to cover the face of the earth, all this complex internal structure has to duplicate, or be doubled exactly. And to do that, the cell must first take into itself an equal amount of smaller molecules ("food") from outside the cell, and rearrange their trillion atoms just right, without error, to make the exact duplicates of its own molecular structure.

How do molecules get duplicated?

The Current (Bankrupt) Theory of How Life Works

There is one central theory or model which is the keystone for giving coherence and "understanding" for the half million researchers in the life sciences. As a model, however, it has been so stretched out of shape to fit the facts that it would be ludicrous, were not most people so blind to it. As a result of this myopia, hundreds of millions of dollars of research money are being wasted in the pursuit of false leads, down blind alleys, while the search for answers to those five basic mysteries we mentioned earlier is postponed.

First, we'll describe as simply as possible this view of how life is supposed to work. Then, we'll discuss just a few examples of how impossible it is, and list over 60 contradictions to it from the research literature. And finally, we'll give a brief description of a totally different alternative explanation of how life really happens—one that is "electromagnetic" rather than "chemical."

Back in the early days, it was discovered that in living cells there were "enzymes" (actually large protein molecules) that acted as catalysts. By their use, many reactions that would either not normally occur at all, or very slowly, could be made to occur quite suddenly. The "enzymes" promoted the reaction without themselves being changed in the process. Little by little the idea

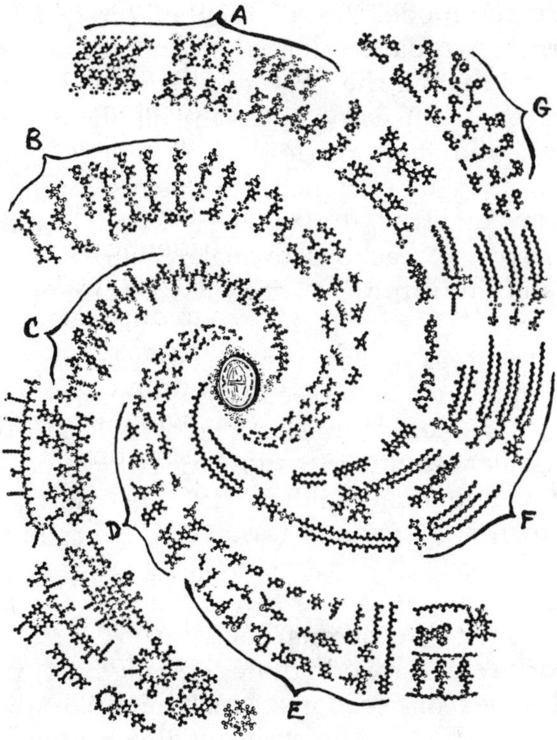

THE MOLECULES OF LIFE These are pictures of the primary bits and pieces which comprise the structure of all life, from virus to man. Some of the groupings are (A) The nucleoproteins, DNA and RNA; (B) Co-enzymes; (C) The amino acids of protein; (D) Sugars and carbohydrates; (E) The vitamins; (F) Lipids (fats); (G) Some typical drugs—not life. These molecules, in various complex combinations, go to make the large visible structures such as we illustrate in Figure 21 —the inside of a tail of a bacterium.

grew that this was the very essence of life, that life was chemical reactions, each mediated by a different catalyst that had just one special job to do.

Thus, the central theory which evolved has three elements: life is "chemical," it consists of reactions in assembly-line fashion, and each reaction is governed by a different enzyme catalyst.

According to this model, therefore, the secret of life could be reduced to three words: sequential enzyme catalysis.

This means that all the large molecules which make up the structure of every living cell (such as the highly ordered proteins, carbohydrates, and fats) are formed in assembly-line fashion, wherein thousands of enzymes are busily zipping atoms and groups of atoms together. It may take a dozen, or a thousand, or millions of stepwise reactions to make them, but that is what makes cells and, in turn, the various tissues and organs of living beings.

The same process was envisioned as embracing the other life functions as well, such as molecular duplication for cell division and growth, for biosynthesis, metabolism, sensitivity, brain function, muscular activity, ingestion, digestion, and excretion, as well as the continuous buildup and breakdown of internal recycling known as "dynamic turnover." Everything was governed by "enzymes."

But surely, this was all checked and doublechecked before it became established as a dogma?

The theory was "proved" quite logically by 1) postulating a hypothetical series of reactions which might lead from some "precursor" molecule to the more complex molecule, and 2) finding among the various cellular extracts an "enzyme" which would catalyze each "step." They never realized that the paralleling of such sequential reactions in the laboratory might be nothing more than artifactual!

But before examining the problems with the theory, let's get more graphic. Imagine a cell magnified to the size of Yankee Stadium. Atoms would be the size of marbles, and the enzymes would be the size of go-carts, moving around at high speed (if their theory is correct), zipping marbles together.

Because there are no gaps, or vacuum, imagine the stadium filled with marbles to the brim, and try to visualize the go-carts speeding through the "solid" mass of marbles, doing their zipping jobs without the guidance of traffic cops. Also note that each zipping job (reaction) involves the approach of enzyme and substrate to an appropriate positioning, a union of the two, the reaction, and then a separating motion, with the enzyme (go-cart) getting out of the way so that the next reaction in the series can be zipped by another and different enzyme next in the sequence—in the next one hundred-thousandth of a second! Whenever a cell divides to become two cells, millions of molecules must be duplicated by zipping together trillions of atoms (marbles)—just right, without error, in incredibly short time periods.

So much for a brief structural description of the theory which holds the life sciences together—so far.

How does the model hold up?

The Many Contradictions to this View

In the accompanying illustration, we have listed over 60 areas in which experimental data contradict the keystone model (sequential enzyme catalysis).

The central theoretical model of modern Biochemistry—STEPWISE ENZYME CATALYSIS—on which the life sciences hang—is contradicted in over 60 areas of experimental data:

Some Problems/Contradictions For Current Theories/Models

1. ENZYME CATALYSIS—no acceptable theory, total mystery . . . many contradictions.

   Enzymes multi-functional—*J. Biol. Chem.* 248, 4144

   pH Optima in Enzyme Series—acid/base switches. see biosynthesis of pyrimidines.

   Problem of diffusion kinetics between reaction sites. *An, Rev. Biochem.* 1970p70. *Science* 172, 185 (Enzyme cluster distances 80 angstroms between sites)

   Bi-functional enzymes, single protein with 2 different catalytic functions. *J. Biol. Chem.* 248, 4144

   Membrane lacks enzymes necessary for synthesis of its phosphatidylcholine lipid. *J. Biol. Chem.* 238, 2615

   150 Restriction Enzymes with 50 different cleavage sites, functions of which vary with cations and ionic conditions. *Science* 205, 455.

   Some enzymes have active and inactive forms.

   Enzyme Systems not sequential, scattered in disarray, divided among structures and Chromosomes. E.J. Gardner. *Princ. of Genetics.* 1966, Wiley. re Histidine operon.

2. DYNAMIC TURNOVER—a horrendous contradiction to stepwise catalysis—the keystone.

All Enzymes have Half-Lives. Thraser, "Turnover of Intracellular Proteins" in *Cellular & Molec, Renewal In Mammalian Bodies,* 1971.

Random Amino acid incorporations in proteins. *J. Biol. Chem.* 247, 5382. *Sci. Am.* Oct. '70.

Left-handed Amino Acids: should be 50/50 mix, but.

Constant renewal of visual cells : labeled amino acids migrate up through rods. *Sci. Am.* Oct. 70.

3. PROTEIN BIOSYNTHESIS AND REPLICATION—embody hopeless problems

Impossible speeds of biosynthesis. *Prog. Biophys. & Molec. Biol.* 18, 285 (1968) etc. etc.

Protein synthesis without nucleotides. *Science* 163, 352. *New Scientist* 10 Oct. 68.

Replication over Whole Membrane Surface. *J. Biol. Chem.* 247, 1180. *J. Cell Biol.* 55, 32

Cyclic Decapeptides: Synthesis without ribosomes or mRNA. *Science* 163, 352

Amino Acids incorporated into Membranes. *J. Biol. Chem.* 247, No. 17, 5382-5390.

Scrappie, self-replicating, without nucleic acids. *New Scientist* 10 Oct. 68.

Effects of Antibiotics on Membrane Protein Syntheses. *J. Molec Biol.* 79(2) 373-389.

Amino acids incorporated into interior of proteins. *Science* 191, 429

Mitochondrial proteins—coding and transport problems. *Science* 189, 789

140 Amino Acids: only 20 specified in genetic code; genetic problem of post-translational modifications for great many structures found in proteins. *Science* 198, 890

Peptidoglycans, Oligopeptides and other amino acid associations; synthesized without ribosomes or nucleoproteins. *Science* 198, 890

Selection of correct aminoacyl-tRNA in protein synthesis, big problem. *Science* 200, 305

4. THE GENETIC CODE AND THE GENE ITSELF—shot full of holes

One Critical Ambiguity (PolyU codes for *both* Phe & Leu)- makes life impossible per code. Other anomalies, such as 'errors' in incorporations, percentages, etc..

Amino acid frequency contrary to what would be expected from the code. *Science* 189, 50

Ribosome genes: scattered between nucleus and plastids (Chlam-monas) *Science* 188, 891.

Enzyme Series scattered on Chromosome: *Science* 192, 55. *Science* 191, 429
Science 188, 891 *B&B Acta* 335 (1974) 396-407.

Human & Chimpanzee Genes: slight differences, many identical. *Science* 188, 107

High frequency of "jumping genes": *Science* 193, 392

Genes for enzymes of E-M glycolytic pathway scattered on 7 chromosomes. *Science* 192, 55

Fracion I Protein: reg. by both nuclear & chloroplast genomes. *Nature* 242, 526. *Sci.* 191, 429.

Ribosomal RNA is not the genetic message. a cheap filler? *Science* 188, 723

Split Genes: many intervening sequences of nonsense DNA. Science 204, 264

Overlapping Genes: a total catastrophe to the gene theory. *New Scientist* 21 Oct 78 *Science* 196, 1187 *Science* 200, 494. etc. etc. looks universal.

mRNA from 5 Different Genes—have same "leader" stretch, 150 nucleotides, located in 3 separate pieces of viral DNA molecule, a distance from any of genes. *Sci. News* 112, 70.

Promoter Sites disappear: deleting first ⅓ and last ⅓ of gene—still produced RNA transcr. *Sci. News Letter* report of Toronto Xlth Int'l. Congress of Biochem.

Interferon Gene: Three separate chromosomes carry same gene. *Science*

Start-Stop Signals for Genes: Not at either end, deep in interior. *Science* 206, 316

Mitochondrial Genetic Code Different: UGA-triptophan instead of stop. CUA-threonine instead of leucine. *Science News* 116, 185. 15 Sep. 1979.

5. DNA REPLICATION—impossible by current models

Watson-Crick model a fiction: with 9 contradictions with experimental facts. *Chem.* Oct. 74p22

Many DNA reaction problems: Erwin Chargoff despairs at ever understanding DNA replication: many reactions mutually exclusive. *Prog. Nucl Ac Res & Mol Biol.* 1976.p1-24.

DNA Replication in Both Directions throughout nucleus, with an unwinding problem. *An. Rev. Genetics* 9, 335 (1975)

PolyA and PolyU Templates for each other. *Science* 192, 633.

Protein Synthesis from DNA w/o RNA intermediate. *Science,* Oct. 1, 1965, 1378

Many RNA molecules in nucleus not involved in protein synthesis. *Cell* 12, 939.

Duplex DNA has 2.9 A' rise, versus Watson -Crick 3.4A". *Science* 201, 525

Enzymes transcribe RNA from RNA templates in various plant species; will accept any RNA single strand. Product a double-stranded RNA. *Science* 205, 859

New DNA-N-Glycolases Enzymes, found in large amounts, remove uracil, 3-M-Adenine, and hypoxanthine, without disturbing sugar-phosphate backbone. *Nature* 278, 121.

6. RIBOSOME ("protein factory") *MODEL*—hopelessly top-heavy model, breaking down.

Ribosome model breaking down; over 100 proteins postulated (more than its own mass) to fulfill its functions. Kurland, *An. Rev. Biochem.* 1971 p377.

Ribosome RNA (⅔ of its mass) non-coding! *Science* 188, 723

Distrib. of Aminoacyl-tRNA Synthetases a problem. *J. Molec. Biol.* 79(2)285.

Polyribosomes in Nuclei (where they don't belong): *Biochem & Biophys.* Sep. 1977, p228

Ribosome Structure mystery: why the same after billion yrs. evolution and drastic changes in proteins—no structural homology between bact. & mammals. . .

7. THE ANTIBODY PROBLEM—still confounding, but can't give up on it.

The Antibody Problem, violates genetic principles. *An. Rev. Biochem.* 1969p415

Immune response: selective binding of Hapten and Antibody cannot be rationalized by electrostatic force calculations. *Nature* 242, 526

Major Histocompatibility Complex (MHC) Antigens and their Genes: confusing subject with many problems, contradictions, and its polymorphism. *Science* 203, 516

8. BASIC BIO-ENERGETICS—hopelessly baffling.

Energy Coupling and Charge Separation mysteries. Green & Ji in *Molec. Basis of Electron Transport*, Academic Press, 1972. Still mysteries in 1980.

High lability of Protons: playing musical chairs at high speeds in macromolecules. *Adv. Biol Med Physics* 1970p3 and *J. Biol. Chem.* 248, 4852 (1973)

Energetics of Virus squirting its DNA into bacillus leaving shell behind . . . what forces?

Superoxide Radical scavenging by family of enzymes called dismutases. artifactual based on chemical assumptions? *Science* 201, 875

Polypeptide hormones (at least 7 or 8) Why do they enter cells? How do they get in? What are they doing there anyway? Mysteries. *Science* 201, 895

We can't begin to go into these, most of them are too technical for the general reader. But we can say something about just two problems which are sufficient by themselves to destroy completely the model. First is the known speeds of some of the events.

Take, for instance, a certain virus (RNA-phage) which, when introduced into the body of a bacterium *(E.Coli)*, duplicates into some 30,000 copies in about 37 minutes. Now this represents an end product of 600 million amino acids and 120 million nucleotides that were synthesized (zipped together). During the last three to five minutes of doubling, this means at least 40 million enzyme-catalyzed reactions per second! These are not mass actions. Each is unique, and sequential, if the model were true.

Going back to our Yankee Stadium model, with the enzymes (go-carts) zipping atoms (marbles) together, each zipping job, with all its coming-togethers, positionings, and pulling-aparts, takes place in a tiny fraction of a second. And the next zap takes place in the next micro-moment, and so on. Imagine thousands of go-carts criss-crossing one square inch in a second! Imagine each of them doing its own unique and different catalytic job and getting out of the way for the next. No traffic cops. All moving at random! Each go-cart somehow doing its correct job at the correct micro-second. All in a solid mass of water and other molecules filling all spaces. Sound kind of crazy? You're right!

But that is what all the experts are teaching! With straight faces! We have many cases of exact knowledge of exact speeds at which the "biosynthetic" zipping-togethers are happening. And right on the face of it, it is absolutely impossible for them to occur according to the sequential model. In our illustration on page 319 we show three Enzymes A, B, C, that are supposed to do zippings 1, 2, 3, 4, 5, involving products a, b, c, d, e, f, g, h, i. Most chemists never "see" the relative sizes and distances of the molecules and processes in which they are dealing, and hence never realize how ridiculous the theory is.

Now, if speeds were not bad enough, there is another factor which slams the door against the sequential model. It's called "dynamic turnover."

Although its discovery was in 1932, there is still no theory to account for the fact that all molecules in life, including all enzymes, are unstable and impermanent, playing "musical chairs," so to speak, and apparently undergoing a continuous process of breakdown and buildup. They have "half-lives," or time periods in which half their atoms are new atoms, ranging from 17 seconds for plasma lipids, to 11 minutes for an enzyme called ornithine decarboxylase, two minutes for some messenger RNA molecules, and up to 300 days for some brain lipids. Whole organelles turn over, with liver ribosomes having a half-life of 5 days, and mitochondria 11 days.

If all those enzymes (go-carts) are spending most of their time in and out of the pits, getting new tires, replacing their nuts and bolts and axles—how can they possibly be doing their unique and specific jobs at the right micro-seconds, while being involved in dynamic turnover?

The fantastic speeds and dynamic turnover are only the tip of the iceberg. Scores of other contradictions add up to an overkill that makes the central theoretical model belief- structure of the life sciences absolutely ridiculous.

No wonder the mysteries of life remain mysteries!

Now let's take a closer look inside living cells.

A Complete Disproof—Illustrated Textbook Cell Chemistry

A, B, and C represent three different "Enzymes" (protein molecules which are catalysts).

1, 2, 3, 4, 5 represent a series of chemical reactions believed to be taking place throughout life—"glycolysis."

a, b, c, d, e, f, g, h, i. represent the small molecules involved in glycolysis.

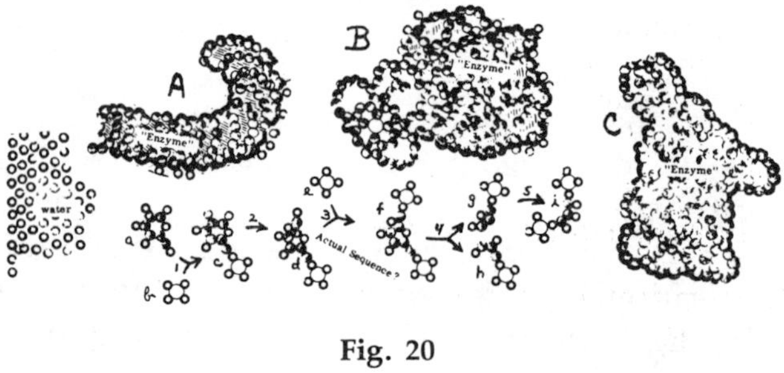

Fig. 20

According to current Biochemistry, the following takes place:

1) Enzyme A triggers a reaction (1) between molecules (a) and (b) to form (c).
2) Enzyme B triggers a chemical transformation of (c) into (d)
3) Enzyme C triggers a reaction (3) between (d) and (e) to form product (f).
4) Enzyme X triggers a splitting of (f) into two molecules (g) and (h).
5) Enzyme Z triggers a reaction between (g) and a PO_4 (not shown) into product (i).

These five reactions must take place in exactly that order, as each product becomes a "substrate" for the next reaction in the sequence. NOW CONSIDER THE TIMING: It all takes place, step-by-step, in a few micro-seconds!

THIS IS MANIFESTLY IMPOSSIBLE BECAUSE:

1) There's no traffic cop to guide (a) and (b) to the exact right position on (A) for the first reaction to take place. Buffeted by the "heat dance" the various (a's) and (b's) move around in random fashion, in every which direction, un-guided.
   It might take seconds or minutes for an (a) and a (b) to just happen to come together—not only at the right place on the right enzyme, but at the right juxtaposition between them. That statistical happening of three bodies coming together rightly—is a very low order of probability.
2) Any one (or all three) of the enzymes are "out to lunch"! Each one of them is involved in a continuous "dynamic turnover" in which they are being broken down and rebuilt . . . and most of the time they are non-functional. They have half-lives of minutes, hours or days, in which they are undergoing turnover.
3) This does not take place in a vacuum. All the spaces between the molecules and the enzymes is filled "solid" with water molecules and other substrates—all dancing the "heat dance," jiggling and jiving, jumping back and forth every which way, getting in each others' ways, bumping and colliding.
4) There are millions of *other* reactions taking place, with thousands of *other* enzymes which are so big that they can't be expected to move around very much or far. The substrates have to seek them out, in a completely random fashion. And every time one product wanders off in the wrong direction, that's the end of that series of reactions—and the result is zero.

The Micro-Structure of Life

As we mentioned earlier, no one knew, a hundred years ago, that the fine (ultra) structure of life was so intricate. At that time it looked like a thick "protoplasm" soup. Hence, it was quite reasonable to settle on a chemical model for how life was happening: chemical reactions, triggered by "enzymes," occurring in random and statistical fashion in the "soup."

But now we have electron microscopes which are getting a sharper and sharper focus on the actual geometrical configurations of the molecules themselves and how the atoms in them are arranged.

What do we see? Instead of protoplasm we find an incredible microcosm of fine structure!

The cell, like a laminar-walled sac, is full of micro-coils, millions of spiral molecules consisting of long strip-sheets of molecular rings, doublets and triplets of which are coiled and recoiled into micro-tubules. The structures have three-dimensional paracrystalline order, with various ring-planes at right angles to other ring-planes, with fine interlaminar connections, with bilateral and other symmetries, and sometimes an overall magnetic configuration. All this is by way of saying that the micro-structure is certainly *not* what could be expected of a random chemical soup but, on the contrary, what is blatantly "electromagnetic."

The molecules have extremely high "dipole moments" (polarity), and are largely "unsaturated" (favoring electrical conductivity), and, given all the spiral structures, and with the continuous electrical activity which exists within living cells (however triggered), it is elementary physics to *expect* magnetic fields not only with respect to the individual atom-rings (perpendicular to the ring-planes and through their centers) but also with respect to the solenoid coils as a whole (with intense longitudinal forces through their axes). The fine structure is "pregnant" with electromagnetic activity.

Fixated on chemical models, researchers have rarely speculated on the micro-physics of life. They have studied magnetic moments, polarizations, electric charges, electrostatic fields, forces and interactions of molecules, and investigated the effects of external magnetic fields on cell functions, but have always stopped short of seeing magnetic fields intrinsic to these activated electromagnetic configurations. One big problem was "charge transfer." How did electrons move through these large molecular structures? Various theories were proposed: band models, hopping models, barrier tunneling, overlapping *pi* orbitals, photoconductors, charge carriers, and ion mobilities, etc. But, somehow, simple electric "currents" with their magnetic fields are shunned almost as a taboo.

How Life Is Really Happening

We're saying the chemical model of life is bankrupt. It hasn't solved any of the fundamental mysteries (*Science* 181:583. We point to an incredibly complex micro-structure and say that what is happening is really electric currents and magnetic fields! You wouldn't ask a chemist to describe how a radio, telephone, or doorbell works, would you? So let's ask a physicist.

Let's start with bacteria and their tails.

For years, chemists thought that bacterial tails (flagella) were whiplashing like little muscles to propel the microbes along in search of food or whatever. By ingenious experiments, Drs. Berg, Silverman, and Simon proved that the tails were actually rotating like propellers (*Scientific American*, Aug '75, 36-44). Because, with the tails stuck to the glass slides, it could be seen that their bodies were spinning instead!

What does the chemist make of this?

They talk about the translocation of ions interacting with fixed charges in the membrane surrounding the tailpiece. But that's insane! According to chemical theory it would take millions of reactions per second to move that tail, in atom-by-atom distances, jerking it around, step by step. The tail, however, is turning at about 3,000 rpm, and smoothly! (*Nature*, 254:389)

How would a physicist view this?

He would think about rotors and stators (the two main parts of an electromagnetic motor) and start looking for windings, electric currents, and magnetic fields to drive the rotor around. And he would wonder whether there might be a commutator in the center of the rotor. A commutator is needed to reverse the magnetic field each half-turn so that the turning can be continuous. But who would imagine a tiny, molecular commutator in the middle of a flagella?

But it's there! At least in the flagella of larger cells.

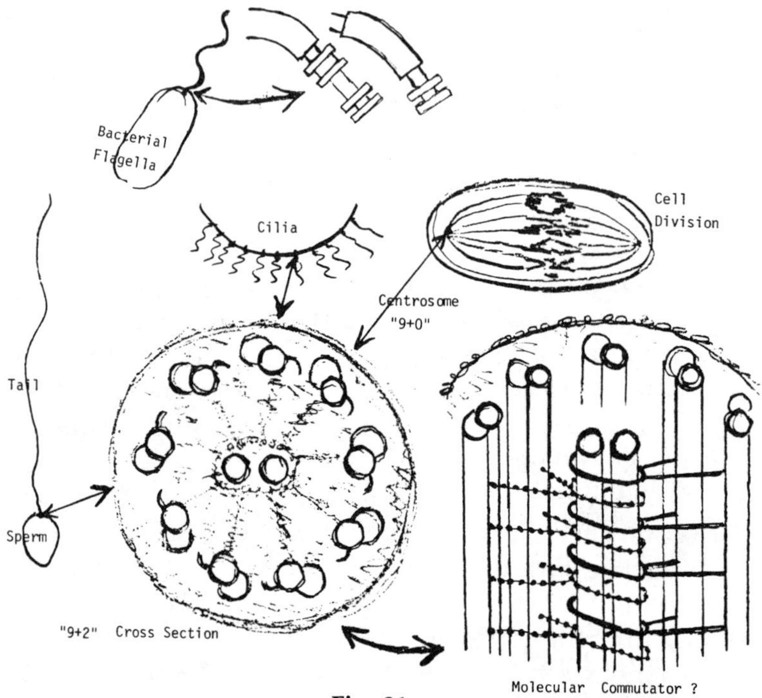

Fig. 21

THE ELECTROMAGNETICS OF LIFE The bacterial tail that spins like a propeller at 3,000 rpm, the "commutator" in the center of a sperm tail, and all the similar micro-structures within cilia and centrioles guiding cell division—all proclaim they are not products of random chemical reactions. Throughout life, the ultrafine electric currents and magnetic fields guide the infinite number of pattern duplications and motions.

In Figure 21, I have sketched the insides of a sperm tail (*Jourmal of Ultra Structure* 33,451-462), to show some of the fine structure which shouts that life is, in fact, electromagnetic. The spinning flagella are clear evidence that the chemical way of thinking about life is dead. I will personally place this picture on the tombstone of the Sciences of Biochemistry.

Of course, there is a sizable area in which chemistry and physics overlap, and of course there are many simple chemical events happening in the blood and guts, etc. But what we are declaring is that the *essence* of life, the basic processes by which it happens, is *not* "chemical." And until the BioSciences as a whole shift over to micro-physics (with its magnetics), they will remain ignorant of the incredible marvel we are.

When, for example, a virus multiplies into tens of thousands of copies in minutes, it is not a chance process of random chemical reactions. There are no maybes. If undisturbed by external forces such as cosmic radiation, the duplication is precise and perfect, down to the last atom, trillions of times without error—over millions of years!

It is not our purpose in this Appendix to cover the new science of life. All we wish to do is to show that Psycles will support, and be supported by, a revolutionary view of life in which its micro-structure declares itself to be "electromagnetic" rather than "chemical."

And in summary, we will close with one picture which illustrates the ultimate secret of life at the micro-level. We portray what happens when the microscopic electric currents spin up and down molecular spirals and around molecular rings, to result in ring-magnetic fields that guide atoms in and out of duplicating patterns. This picture is at the unit ring level. It shows the atoms of just one ring of long, multi-ring chains which, in turn, coil and re-coil into larger molecular structures.

It's incredibly simple!

With this one ring-structure, electromagnetically activated, we can *see* all the fundamental secrets of life in action: how molecules duplicate, how they interact, how they build up and break down continuously in "dynamic turnover," the secret of catalysis, the secret of motility—the attractions and repulsions of muscular activity, the secret of homeostasis, and the whole balance manifest throughout the fine processes of life.

Look at it!

With this illustration of the electromagnetic principles at the unit-ring level, we can *see* the electrons spinning around the ring, *see* the magnetic fields at right angles to the ring and through its center, *see* the attractions and repulsions of atoms and small groups of atoms (the "substrates" in the whole recycling process), *see* the duplication, *see* the motilities, *see* the balance. . . .

Earlier we saw the same magnetics at the holistic level of our Psycles philosophy. Here it is at the micro-level.

It is an electromagnetic event at all levels.

Simple, beautiful, orderly, balanced, errorless.
That is the way life is.
That is the way we are.

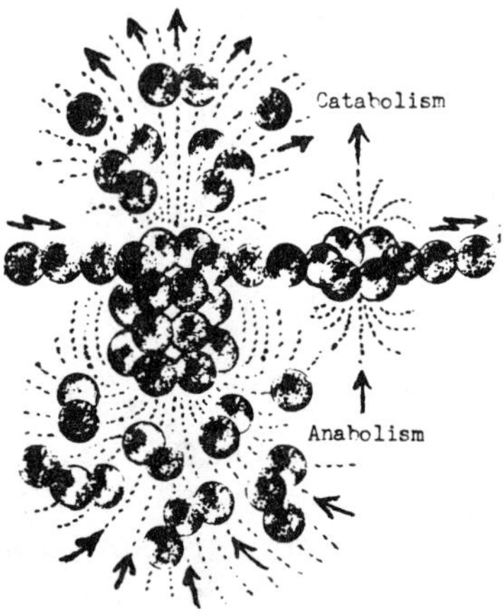

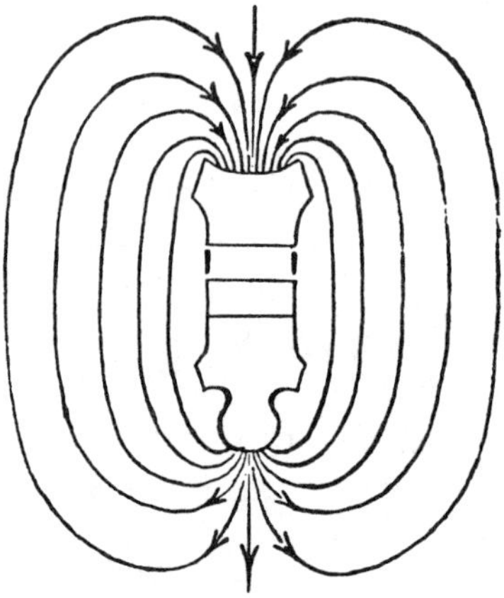

Science and Religion

Thus, carrying it right down to the micro-level, we can see that the missing ingredient in the age-old conflict between science and religion has been the unifying magnetic field.

Without magnetics, we have the hard-line scientists and sociobiologists led by Kurtz, Skinner, Sagan, and Asimov, who are vehemently opposed to "psi" phenomena. Immersed in quantum mechanics and seeing everything as particles, they cannot comprehend subconscious awareness and communication, ESP, the potentials for perfect health, the global omniscience, nor the Unitive Reality which enfolds all. Theirs is the "Robot View," the bottom half of reality, which has been expressed most eloquently by Jacques Monod in his book *Chance and Necessity*. According to him, all life

> "is a product of an enormous lottery presided over by natural selection, blindly picking the rare winners from among numbers drawn at utter random. . . . Man knows, at last, that he is alone in the universe's unfeeling immensity, out of which he emerged only by chance."

That is the "Robot View."

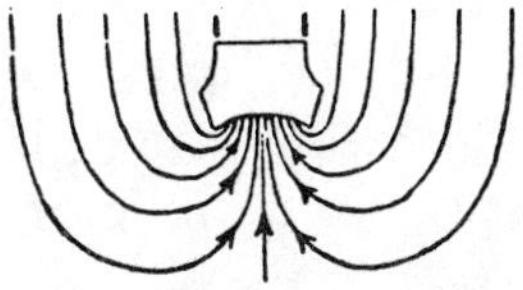

But the other half of the picture is equally false. Sometimes called "Idealists," there are those like Kant and Berkeley, to whom reality is nothing but a "head trip." A recent example comes from *Seth,* author Jane Morgan's masculine alter-ego. (S)he states that all reality is generated out of your head. If six people sit looking at a cup in their center, there are six minds creating, *de novo,* six different physical cups. This is nonsense. The ultramystical, purely psychic premises can be just as absurd as the pure "mechanism" of Sagan, Asimov, and Monod's "Robot View."

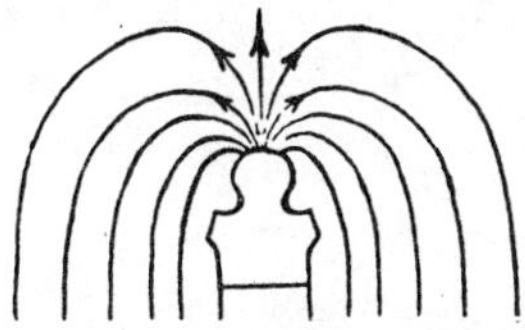

With Psycles, and the overall BioMagnetic philosophy, we have for the first time a binding link which unifies both the heads and tails to reveal the true holism which exists. It is a Unitive Philosophy. It does not just incorporate the mechanistic facts of science. Many have tried that. It sees them in a new light.

Yes, there are these physical structures and forces. Yes, we can measure their stimulus-response parameters. But we *also* see our central role in generating these outcomes—through intentions, based on identifications. We see the events not just as they reenter, so to speak, at the anal end. We also see their genesis out of the top of the head, and coming around!

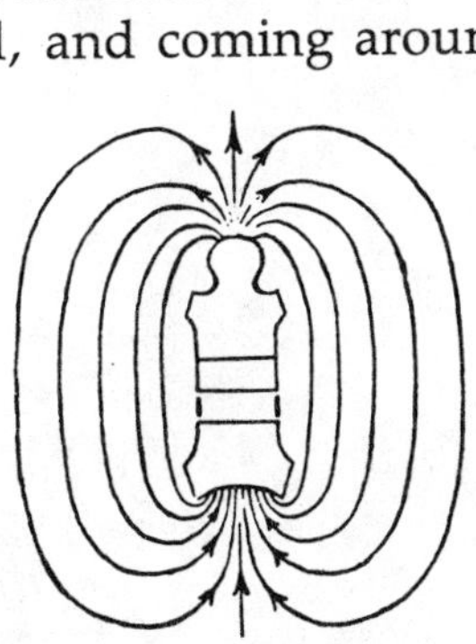

This being the case, we see how things can be changed: by changing images and identities in the center! Psycles doesn't just bridge. It unifies science and religion with a whole new comprehension of the very process!

With this new vision of life, we can embark on a new adventure to claim our birthright.

We can transform our very existence.

All it takes is the decision to do so.

We all know how. The whole universe is ready to help you.

Now go for it!

Index